Food Selection and Preparation

A Laboratory Manual

SECOND EDITION

Food Selection and Preparation

A *Laboratory Manual*

SECOND EDITION

Frank D. Conforti

WILEY-BLACKWELL

A John Wiley & Sons, Ltd., Publication

Second edition first published 2008
© 2008 Frank D. Conforti
First edition first published 1997
© 1997 Iowa State University Press

Blackwell Publishing was acquired by John Wiley & Sons in February 2007. Blackwell's publishing programme
has been merged with Wiley's global Scientific, Technical, and Medical business to form Wiley-Blackwell.

Editorial offices:

2121 State Avenue, Ames, Iowa 50014-8300, USA

For details of our global editorial offices, for customer services and for information about how to apply for
permission to reuse the copyright material in this book please see our website at
www.wiley.com/wiley-blackwell

Library of Congress Cataloging-in-Publication Data

Conforti, Frank D.
    Food selection and preparation : a laboratory manual / Frank D.
Conforti. – 2nd ed.
        p. cm.
ISBN-13: 978-0-8138-1488-9 (alk. paper)
ISBN-10: 0-8138-1488-X (alk. paper)
1. Food–Laboratory manuals.   2. Cookery–Laboratory manuals.   I. Title.
    TX354.C64 2008
        664–dc22   2008029048

A catalogue record for this book is available from the British Library.

Set in 11/12 pt ArialNarrow by Aptara® Inc., New Delhi, India
Printed in Singapore by Fabulous Printers Pte. Ltd.

1   2008

Figure 4.3 was adapted from General Mills, Inc., *Betty Crocker's Cookbook*, copyright 1978, pg 213; Figures
8.1, 8.2, and 8.3 were adapted from General Mills, Inc., *Betty Crocker's Microwave Cookbook*, Random
House, Inc., 1981; all figures appearing in Appendix H were adapted from General Mills, Inc., *Betty Crocker's
Best Recipes for Meat and Vegetable*; the Chicken Foodservice cuts chart in Appendix G is reprinted from
North American Meat Processors, Chicken Foodservice Poster, Revised May 2006, reprinted with permission
of John Wiley & Sons, Inc.; and the Pork Foodservice Cuts chart in Appendix G is reprinted from North
American Meat Processors, Pork Notebook Guides, Revised May 2006, reprinted with permission of John
Wiley & Sons, Inc.

Contents

Preface

My objective in writing this manual was to create a learning tool for the student in food service, hospitality management, dietetics, or family and consumer education. Ten years have elapsed since the first edition of this manual. During those 10 years many changes have taken place in food selection and preparation. In order to keep up with the changing tide in food, this manual has been updated to reflect these current trends. There are new recipes and reformulation of existing recipes with regard to ingredients or manipulation. There are also updates of technical information in each unit to be in touch with the current trends and discoveries in food.

The student should learn how to prepare nutritious food and how to make substitutions when necessary, yet still maintain the integrity and quality of food. Therefore, the student must understand the function of the ingredient(s) in a particular food system. The student must understand why the ingredient is being added and what effect the ingredient will have on the quality of the food product during preparation. This manual hopes to carry out this purpose not only with the recipes that are found in each particular unit, but also the questions, exercises, and vocabulary words that are part of each unit.

Each laboratory is an independent unit and can be assigned according to any sequence chosen by the instructor. There are a number of recipes in each unit, but they all do not have to be included in the lesson especially if some laboratory periods run for 2 hours instead of 3 hours. A careful selection of activities by the instructor should give the student a firm basis in foods and a clear understanding of the proper selection and manipulation of ingredients that will lead to a quality product.

I hope that you will enjoy this manual as much as I have over the years in the development of its content. I have to give credit to the students (and there have been over 2,000 since the first edition had come out!) who have contributed to its success. It is because of these students' incisive recommendations, constructive criticisms, and devotion to the subject of food that this manual has evolved into what it is today. It is hoped that many more students will become acquainted with its contents, and that they will come away with an interest and deep respect for food and the contribution that food makes in one's health and daily life. Finally, I hope that this manual will make a contribution by being a continuing source of information long after the course is completed.

Acknowledgments

A revision of a book takes time, patience, and the support of many people. I would like to express my appreciation to the following people who have contributed to the revision of this manual: Sherry Seville, Virginia Tech, whose expertise at the computer assisted in formatting the revised manuscript for publication; Sharon Kast, also of Virginia Tech, whose time and patience were responsible for the photographs that appear in some of the laboratory units; and especially to the students whose suggestions and participation over the years have made this manual an integral part of the Food course at the Department of Human Nutrition, Foods, and Exercise.

LABORATORY 1

Measuring Techniques

LABORATORY 1
Measuring Techniques

Proper measuring techniques must be emphasized to ensure success in food preparation. There are differences when measuring liquid and dry ingredients, and the student must learn these techniques as soon as possible in order to be successful in food preparation. The objective of this laboratory exercise is to introduce the student to proper measuring techniques.

VOCABULARY

boiling point	meniscus	solvent
conduction heat	opaque	solute
convection heat	simmering	

MEASURING TECHNIQUES

The American Standards Association has defined the capacities of various measures, but not all measuring equipment has been standardized to meet these specifications. Variations of 5%, more or less than standard, are allowable.

I. NONMETRIC MEASURE OF VOLUME

A. DRY MEASURES

A set of dry measuring cups includes measures for 1/4 cup, 1/3 cup, 1/2 cup, and 1 cup (there are some manufacturers that make 2/3 cup and 3/4 cup measures). These measures are used for dry ingredients and solid fats. Ingredients vary in the way they pack down, lump, or cling to the measuring cup. Use the following guidelines when measuring:

1. All-purpose flour, cake flour, granulated sugar, and confectioner's sugar should be lightly spooned into the appropriate size dry measuring cup. **Do not shake or pat down**. Use a straight-edged spatula or knife to level off ingredients (Fig. 1.1).

FIG. 1.1: Spoon dry ingredients lightly into cup and level off with a straight-edged spatula.

2. Nuts, coconut, and bread crumbs should be spooned into the cup and packed down lightly.
3. Brown sugar should be spooned into the dry measure cup and packed down firmly with spatula and spoon.
4. Solid fats include hydrogenated shortening, lard, margarine, and butter. The solid fat should be packed into the dry measure with firm pressure. Butter and margarine should be at room temperature before being measured.

B. SMALL AMOUNTS OF INGREDIENTS

1. Baking powder, baking soda, salt, and spices are used in such small amounts that they must be measured in small capacity measures of 1 tablespoon or less.
2. Ingredients should be stirred and free of lumps.
3. The desired measure is dipped into the ingredient and leveled off.
4. Usually, the measuring spoons are found as 1/8 teaspoon, 1/4 teaspoon, 1/2 teaspoon, 1 teaspoon, 1/2 tablespoon, and 1 tablespoon.

C. LIQUIDS

1. Oil, honey, milk, molasses, water, melted fat, and other liquid ingredients should be measured in a graduated, transparent liquid measure with a pour spout.
2. Fill the measure to the desired graduation and check it by holding the measure at eye level so the bottom of the meniscus—the curved, upper surface of the liquid—matches the desired line on the side of the measure (Fig. 1.2).
3. Opaque liquids (such as milk and honey) that do not show a meniscus are measured by aligning the top of the liquid with the line on the measure.
4. Many liquids, especially oil and honey, tend to cling to the sides of the cup. To obtain an accurate transfer of the liquid, it is essential that the inside of the cup be scraped out with a rubber spatula. **Hint: spray measuring cup with cooking spray before measuring molasses or honey. This will make removal of the ingredient more efficient.**

FIG. 1.2: Read the measure by holding it at eye level so the bottom of the meniscus matches the desired line on the side of the measure.

D. OTHER MEASURING ADVICE

1. If the recipe specifies 3 teaspoons of baking powder, the tablespoon measure should be used to make the measurement. To measure 3 separate teaspoons introduces greater error in measurement.
2. When the recipe specifies less than 1 cup of liquid, and the measurement is made in a 2-cup graduated measure, there is also a greater chance of error.
3. It is important to use the measuring utensil that is closest in size to the amount of ingredient for greater accuracy.

EQUIVALENT MEASURES

1 tablespoon	= 3 teaspoons		3/4 cup	= 12 tablespoons
1/8 cup	= 2 tablespoons		1 cup	= 16 tablespoons or 1/2 pint
1/4 cup	= 4 tablespoons		1 pint	= 2 cups
1/3 cup	= 5 tablespoons + 1 teaspoon		1 quart	= 4 cups or 2 pints
1/2 cup	= 8 tablespoons		1 gallon	= 4 quarts
2/3 cup	= 10 tablespoons + 2 teaspoons			

II. TO LEARN CORRECT TECHNIQUES FOR MEASURING INGREDIENTS

A. FLOUR (ALL-PURPOSE OR CAKE)

1. Method 1

    a. Fill 1/2 cup dry measure by dipping into canister of flour.
    b. Level with spatula.
    c. Weigh flour on gram scale and record weight in Table 1.1.
    d. Repeat.

2. Method 2

    a. Place 1/2 cup dry measure on a piece of waxed paper of 12 square inch.
    b. Sift flour directly into the cup until the cup overflows. Do not let the sifter touch the cup.
    c. Level flour with the edge of the spatula.
    d. Weigh flour and record weight in Table 1.1.
    e. Repeat.

3. Method 3

    a. Stir flour in canister to lighten.
    b. Carefully spoon flour 1 tablespoon at a time into 1/2 cup dry measure.
    c. Level flour with the edge of the spatula.
    d. Weigh flour and record weight in Table 1.1.
    e. Repeat.

Table 1.1 EVALUATION OF THE WEIGHT OF 1/2 CUP OF FLOUR			
Method	Trial 1	Trial 2	Standard Weight*
1			
2			
3			

*All purpose: 1/2 cup, sifted: 58.0 g; 1/2 cup, spooned: 62.5 g; cake flour: 1/2 cup, sifted: 48.0 g; 1/2 cup, spooned: 55.5 g.
Source: *Handbook of Food Preparation: Food and Nutrition Section*, 9th edition, 1993, American Home Economics Association, p. 182.

QUESTIONS

1. Which method of measuring flour yields the best check? Why?

2. What would cause a difference in weight from the standard?

3. How would you substitute all-purpose flour for cake flour in a recipe? Would this substitution work for all type of baked products?

4

B. SUGAR: GRANULATED AND BROWN

1. Method 1

   a. Fill a 1/4 cup dry measure with granulated sugar by dipping it into the canister.
   b. Level the sugar with the edge of the spatula.
   c. Weigh sugar and record in Table 1.2.
   d. Repeat.

2. Method 2

   a. Fill a 1/4 cup dry measure with brown sugar by spooning sugar into cup.
   b. Level the sugar with the edge of the spatula.
   c. Weigh sugar and record in Table 1.2.
   d. Repeat.

3. Method 3

   a. Fill a 1/4 cup dry measure with brown sugar by pressing the sugar into the measuring cup.
   b. Level the sugar with the edge of the spatula.
   c. Weigh sugar and record in Table 1.2.
   d. Repeat.

Table 1.2 EVALUATION OF THE WEIGHT OF 1/4 CUP OF SUGAR			
Method	Trial 1	Trial 2	Standard Weight*
1			
2			
3			

*Light brown sugar, packed: 1/4 cup = 50 g; dark brown sugar, packed: 1/4 cup = 50 g; granulated sugar: 1/4 cup = 50 g.
Source: *Handbook of Food Preparation: Food and Nutrition Section*, 9th edition, 1993, American Home Economics Association, p. 195.

QUESTION

1. How does the method for measuring brown sugar differ from that of measuring granulated sugar?

C. LIQUID

1. Method 1

   a. Fill a liquid measuring cup with water to 1/4 cup mark.
   b. Place a cup on a level surface and position yourself at eye level with the water before attempting to read the water level (Fig. 1.2).
   c. Transfer all the water from the measuring cup to a 100-mL graduated cylinder and read the volume in milliliters.
   d. Record the volume in Table 1.3 and repeat.
   e. Repeat Steps a through d, but use milk.

2. **Method 2**

    a. Fill a 1/4 cup dry measure with water.
    b. Place measure on a level surface and position yourself at eye level with the water before reading the water level.
    c. Transfer all the water from the cup to a 100-mL graduated cylinder and read the volume in milliliters.
    d. Record the volume in Table 1.3 and repeat.

Table 1.3 EVALUATION OF LIQUID MEASUREMENTS			
Method of Measurement	Trial 1	Trial 2	Standard Volume*
1			
2			

*1 cup liquid measure = 236 mL; 1/4 cup liquid measure = 59 mL.
Source: *Handbook of Food Preparation: Food and Nutrition Section*, 9th edition, 1993, American Home Economics Association, p. 180.

QUESTIONS

1. Was there a visual difference between the water and milk when they were measured, and why?

2. What is the error that occurs when using a dry measure for measuring liquids?

D. FATS

1. **Method 1**

    a. Fill a 1/4 cup dry measure with a solid fat.
    b. Using a rubber spatula, press fat into the cup making sure there are no air pockets.
    c. Level off with a straight-edged spatula.
    d. Carefully, remove fat from cup with a rubber spatula and weigh.
    e. Record weight in Table 1.4 and repeat.

2. **Method 2**

    a. Melt solid fat in a saucepan over **low heat**.
    b. Take a 1 cup liquid measuring cup and pour melted fat up to the 1/4 cup measure mark.
    c. Weigh and record the weight in Table 1.4 and repeat.

Table 1.4 EVALUATION OF THE WEIGHT OF 1/4 CUP HYDROGENATED FAT			
Method	Trial 1	Trial 2	Standard Measure*
1			
2			

*Hydrogenated shortening, solid: 1/4 cup = 46 g.
Source: *Handbook of Food Preparation: Food and Nutrition Section*, 9th edition, 1993, American Home Economics Association, p. 175.

QUESTIONS

1. What precautions should you take for measuring fats?

2. Account for the differences in weight of the fats.

3. Why is it important to allow fats, such as butter and margarine, to come to room temperature before measuring and mixing?

III. WATER AND THERMOMETRY

1. Most of the changes brought about by foods by cooking take place in a watery medium (moist heat).
2. Water absorbs heat from the hot unit through the cooking utensil and transfers this heat to the food.
3. When water boils, convection heating currents surround the food; therefore, even and quick cooking of the food occurs.
4. Water sets its limit to how hot it gets ($100°C$ or $212°F$), while fat can go to higher extremities.
5. The intensity of the heat is measured by a thermometer (either in $°F$ or $°C$).

A. FACTS ON USING A THERMOMETER

1. The bulb must be completely covered with hot liquid.
2. The bulb should not touch the sides or bottom of the utensil.
3. There are $100°$ between the boiling point and the freezing point of water on the centigrade scale.
4. There are $180°$ between the boiling point and the freezing point of water on the Fahrenheit scale.
5. Therefore,
   a. $1°C = 1.8°F$
   b. $°C = (°F - 32) ÷ 1.8$
   c. $°F = (°C × 1.8) + 32$

B. LEARN TO RECOGNIZE COMMONLY USED TEMPERATURES

Heat water to each temperature specified in Table 1.5 and note its appearance.

Table 1.5 EVALUATION OF COMMONLY USED TEMPERATURES			
Term	Description	°F	°C
Room		77.0	25
Lukewarm		98.6	37
Scalding*		149.0	65
Simmering		185.0	85
Boil slowly		212.0	100
Boil rapidly		212.0	100

*The temperature varies with material being scalded.

QUESTIONS

1. Explain what happens when water boils.

2. Name some instances when scalding temperature is used in food preparation.

3. What happens when salt is added to boiling water? If sugar is added?

C. DETERMINING THE ACCURACY OF LABORATORY OVENS

1. Take an oven thermometer and calibrate your ovens. Place the rack in the middle of the oven. Use 350°F as a standard to go by.
2. Record oven temperature: _____

QUESTIONS

1. Why is it important that the temperature of the oven be exact?

2. In what position would you place the oven rack to cook food in a conventional oven for:

    a. a two-layer cake?

    b. a tube cake pan?

    c. a cookie sheet pan?

    d. a roasted whole turkey?

3. What is a convection oven? What temperature adjustment is made when using such an oven? Is the rack adjustment the same for the products mentioned in Question 2 for the convection oven?

IV. APPLICATION OF MEASURING TECHNIQUES: COOKIES

OBJECTIVES

1. To practice proper measuring techniques involving dry and liquid measuring.
2. To familiarize the student with reading a recipe and becoming acquainted with certain culinary terms.

A. CHOCOLATE CHIP COOKIES

1/3 cup shortening	3/4 cup + 1 tablespoon all-purpose flour
1/4 cup granulated sugar	1/4 teaspoon baking soda
1/4 cup light brown sugar, packed	1/4 teaspoon salt
1 large egg	3 oz. semisweet chocolate chips
1/4 teaspoon vanilla extract	

1. Preheat oven to 375°F. Make sure oven rack is in the middle position.
2. **Sift** together flour, salt, and baking soda; set aside.
3. In a medium-sized bowl, **cream** together shortening, granulated sugar, brown sugar, egg, and vanilla for 2 minutes.
4. Add flour mixture to creamed mixture; **mix** only until flour is combined.
5. **Stir** in chocolate chips. Chill dough in freezer for 5–10 minutes. (This helps the dough from not overspreading during baking.)
6. Drop dough by rounded teaspoonfuls about 2 inches apart onto ungreased baking sheet.
7. Bake for 8–10 minutes, or until edges start to brown slightly. Remove pan from oven, and allow cookies to cool for 2 minutes on the pan.
8. With a spatula remove cookies from pan and place on a wire rack to cool.

B. CHOCOLATE CHIP COOKIES (LOWER-FAT VERSION)

1/2 cup granulated sugar minus 1 tablespoon	1 cup + 2 tablespoons all-purpose flour
1/4 cup light brown sugar, packed	1/2 teaspoon baking soda
1/4 cup butter or margarine at room temperature	1/4 teaspoon salt
1 teaspoon vanilla	1/4 cup miniature semisweet chocolate chips
1 egg white	

1. Preheat oven to 375°F. Make sure that the rack is on the middle position in the oven.
2. Sift together, flour, salt, and baking soda; set aside.
3. In a medium-sized bowl, **cream** together butter, granulated and brown sugar, egg white, and vanilla for 2 minutes.
4. Stir in flour mixture until just combined. Stir in chocolate chips. Chill dough for 5–10 minutes in the freezer.
5. Drop dough by rounded teaspoonfuls about 2 inches apart onto an ungreased baking sheet.
6. Bake for 8–10 minutes, or until edges are lightly brown.
7. Remove pan from oven and allow cookies to cool for 2–3 minutes before removing with a spatula to a cooling rack.

C. OATMEAL COOKIES (BASIC RECIPE)

1/2 cup all-purpose flour	1/2 cup light brown sugar, packed
1/2 teaspoon baking powder	1 large egg
1/4 teaspoon salt	$1\frac{1}{2}$ cups quick cooking oatmeal
3/4 teaspoon cinnamon	1/4 cup chopped walnuts or pecans
1/4 cup + 3 tablespoons milk	1/4 cup sweetened coconut
1/4 cup + 2 tablespoons shortening	1/4 cup chopped raisins or dates

1. Adjust the rack to the middle of the oven. Preheat oven to 375°F.
2. Sift together flour, baking powder, salt, and cinnamon into a medium-sized bowl.
3. Add shortening, brown sugar, milk, and egg to flour mixture and beat until smooth.
4. Add oatmeal and mix thoroughly.
5. Add walnuts, coconut, and raisins and mix until combined.
6. Drop dough by teaspoonfuls onto a greased cookie sheet.
7. Bake for 12–15 minutes. When cookies appear dry and the edges are light brown, remove them from oven. Cool slightly and then remove cookies from the sheet onto a cooling rack.

D. OATMEAL SPICE COOKIES (LOW-FAT VERSION)

2¼ cups quick cooking oatmeal
2 tablespoons orange juice
1 cup all-purpose flour
1/2 teaspoon baking soda
1/2 teaspoon baking powder
1/4 teaspoon salt
1/4 teaspoon cinnamon
1/8 teaspoon cloves

1/8 teaspoon nutmeg
3 tablespoons margarine or butter, softened
3 tablespoons canola oil
3/4 cup dark brown sugar, packed
1 tablespoon molasses
1 large egg white
2 teaspoons vanilla extract
2 tablespoons granulated sugar—for shaping cookies

1. Preheat oven to 350°F. Spray two baking sheets with vegetable spray; set aside. Adjust racks in the oven to be one at the top and the other at the bottom.
2. Stir together oats and orange juice in a medium-sized bowl; set aside.
3. Sift together flour, baking soda, baking powder, salt, cinnamon, cloves, and nutmeg onto a sheet of wax paper; set aside.
4. In a large mixing bowl, with mixer at medium speed, beat margarine and oil until well blended and smooth.
5. Add brown sugar, molasses, egg white, and vanilla and beat until smooth and fluffy.
6. Blend in flour mixture.
7. Using a wooden spoon, mix in oatmeal mixture until thoroughly incorporated.
8. Shape dough into 1 inch balls, and place 3 inches apart onto baking sheets. Flatten cookies using the bottom of a glass that has been lightly greased and dipped in the 2 tablespoons of granulated sugar. Dip the glass in sugar each time after flattening each cookie.
9. Place pan in preheated oven. After 4 minutes, change the position of the baking sheets and bake cookies for another 4–5 minutes.
10. Let cookies stand on sheets for 3–4 minutes. Using a spatula, transfer cookies to cooling racks and let stand until completely cooled.

E. SNICKERDOODLES

1¾ cups all-purpose flour
1/2 teaspoon baking soda
1/2 teaspoon cream of tartar
1/4 teaspoon salt
1 cup minus 1 tablespoon granulated sugar
1/4 cup butter, softened

1 tablespoon light corn syrup
1½ teaspoons vanilla extract
1 large egg
2 tablespoons granulated sugar
1½ teaspoons cinnamon

1. Preheat oven to 375°F. Adjust racks one at the top of oven, the other at the bottom. Spray two baking sheets with cooking spray; set aside.
2. Sift together flour, baking soda, cream of tartar and salt; set aside.
3. Combine sugar (1 cup minus 1 tablespoon) and butter in a large mixing bowl, and beat with a mixer at medium speed until well blended. Add corn syrup, vanilla, and egg; beat well. Gradually, add flour mixture (in three additions) to sugar mixture, beating just until combined. Cover and chill for 10 minutes.
4. Combine 2 tablespoons sugar and cinnamon, and blend well in a small dish.
5. Shape dough into 30 balls. Roll balls in sugar mixture. Place balls 2 inches apart onto baking sheets. Flatten balls with bottom of glass. Bake cookies for 4 minutes; then rotate pans by changing their positions in the oven and bake for another 4 or 5 minutes (cookies will be slightly soft). Cool cookies on baking sheets for 2 minutes. Remove cookies from pans and place on wire cooling racks.

F. BROWNIES

2 oz. unsweetened chocolate
1/3 cup shortening
1 cup granulated sugar
3/4 teaspoon instant coffee granules
2 large eggs

1/2 teaspoon vanilla extract
1/2 cup + 2 tablespoons all-purpose flour
1/2 teaspoon salt
1/2 cup semisweet chocolate chips
1/2 cup walnuts, chopped (optional)

1. Preheat oven to 350°F. Make sure that the rack is in the middle position.
2. Grease an 8 × 8 × 2 cubic inch pan.
3. Melt chocolate and shortening in a 2-quart saucepan over **low heat**. Stir constantly with a wooden spoon. As soon as the mixture has melted, remove it from heat and cool.
4. Mix in sugar, eggs, instant coffee granules, and vanilla.
5. Stir in remaining ingredients. Spread in pan.
6. Bake for 30 minutes or until brownies start to pull away from the sides of the pan (if using a toothpick to test for doneness, the toothpick should have some moist crumb attached to it when removed). **Do not overbake or the brownies will be dry**.
7. Cool slightly. Cut into bars 2 × 1½ inches. Place bars on cooling racks.

GENERAL QUESTIONS

1. How should brown sugar and solid fat be measured? Explain why these ingredients must be measured in this way.

2. What does it mean to "cream?"

3. Why should the oven rack be in the middle position when baking the cookies? Why were the cookies (low-fat oatmeal and snickerdoodles) rotated in the oven mid-way during their baking?

4. If a glass baking dish was used for making the brownies, what change in the baking temperature would there be?

5. It is assumed that a conventional oven is used for the recipes in this unit. What changes in temperature and time would there be if a convection oven was used? How should rack position be addressed in the convection oven?

6. Why were the cookies placed on cooling racks?

7. How important is exact oven temperature to baking quality?

LABORATORY 2

Food Preservation: Canning and Freezing

LABORATORY 2
Food Preservation: Canning and Freezing

The foods we eat should be wholesome, nutritious, and safe. This laboratory exercise will demonstrate to the student how to extend the shelf life of food by using extremes in temperatures, heat (canning), and cold (freezing).

VOCABULARY

antioxidant	headspace	pressure canner process
blanch	high acid food	sterilization
boiling water bath	low acid food	turgor
freezer burn	polyphenoloxidase	venting

OBJECTIVES

1. To examine the effect of extremities in temperature (heat versus cold) in extending the shelf life of food.
2. To demonstrate and observe the differences between raw pack and hot pack.
3. To demonstrate and discuss the differences between the boiling water bath process and the pressure canner process.
4. To prepare jellies, jams, conserves, and pickles as a form of preserving fruits and vegetables.
5. To define and demonstrate the differences between freezing fruits and vegetables as a simple method of preservation.

PRINCIPLES

1. **Time** and **temperature** are very important factors when canning. These two factors ensure that the product will be free of microorganisms and at the same time be sterile for a long shelf life without refrigeration.
2. The canning process (heat treatment) to be used is determined by the pH of the food:
    a. *High acid foods* (pH < 4.6) (fruits, pickles, jams, jellies, and some tomato products) are processed by the **boiling water bath** (212°F).
    b. *Low acid foods* (pH ≥ 4.6) (vegetables, poultry, meat, milk, fish, soups, etc.) are processed by the **pressure canner process** (240°F, at 10 pounds pressure).
3. Food for canning can be packed by either **raw pack** or **hot pack**. The pH of the food does not dictate how the food would be packed in the jar. Raw pack guarantees more food identity, but more food will fit into the jar by the hot pack method.
4. Foods (pH ≥ 4.6) processed under pressure must be done to insure the destruction of *Clostridium botulinum*.
5. Pickling of food involves vinegar (acid) and salt, whereby, the **boiling water bath method** will suffice in lowering the microbial load.
6. When canning food, the **time of processing** is dependent on size of the jar (large versus small); type of jar (glass versus metal); type of food (hard versus soft); type of packing (raw versus hot).
7. Jellies are made by a balanced formulation of fruit, pectin, acid, and sugar.
8. Enzymes are responsible for the darkening of sliced fruits and vegetables. Pretreatment of fruits and vegetables prior to freezing is necessary because freezing only **slows down** the enzyme but **does not destroy** the enzyme.
9. Fruits, because of their texture, cannot be blanched but are treated with an antioxidant prior to freezing.
10. Vegetables are blanched prior to freezing in order to inhibit enzyme action.
11. Containers in which fruits or vegetables are stored during freezing must be tightly sealed in order to prevent freezer burn (sublimation).

I. OBSERVE AND LEARN HOW TO USE UTENSILS AND EQUIPMENT COMMONLY USED IN CANNING

A. JARS AND THEIR CLOSURES

1. Check for and discard any glass jars with cracks or chips and any rings with dents or rust; these defects prevent any airtight seals.

2. Wash jars in **hot soapy water** and **rinse well** (jars and rings are reusable); place jars upside down on a clean cotton towel and do not invert them until you are ready to fill the jar.
3. Place lids in a pan of water; bring to a boil. Remove from heat and leave in hot water until ready to use. This softens the rubber on the lid and provides for an airtight seal.

> *Label each lid*:
> **Food, Pack (if applicable), and Process**
> **Date canned**
> **Laboratory day and time**
> **Kitchen#**

B. METHODS OF PACKING

1. *Raw Pack*: Uncooked food is placed into jars. The food is then covered with boiling liquid, leaving headspace recommended (Fig. 2.1).
2. *Hot Pack*: Food is heated in syrup, water, steam, or extracted juice. The food is packed in jars and covered with boiling liquid, leaving headspace recommended.

FIG. 2.1: Headspace prevents loss of liquid and effects seal in vacuum-type closures.

C. PROCESSING EQUIPMENT

1. *Water Bath*: Select a deep pot with a lid. There should be a rack on the bottom of the pot so that the jars can rest on during processing. There should be enough water in the pot, so when the jars are placed in the pot the water should be 2–3 inches above the jars. The water should be boiling when the jars are added to the pot. As soon as the jars are added the water will stop boiling. Start timing when the water starts to boil.
2. *Pressure Canner*: Put water in the canner to a depth of 2–3 inches (**source of steam for processing**). Place jars on rack in canner. Fasten lid securely. Open the pet cock and place the canner on the heat. When steam starts to escape in a steady stream, start timing for 7–10 minutes. Close the pet cock and allow the pressure to build. Most foods at pH $\geq$ 4.6 are processed at 240°F at 10 pounds pressure. As soon as gauge registers this value, start timing and hold at 240°F for the allotted time. **In order to hold the pressure constant this is achieved by controlling the heat to the canner**. At the end of the processing time, turn off the heat and allow the gauge to return to zero before opening canner and removing jars.

D. COOLING OF PROCESSED JARS

After jars are removed from the processing equipment, they are placed on a clean cloth. Allow some space between the jars for cooling and air circulation. Check the lids after cooling; the center of the lid should be depressed if sealed. If the lids "pop" up and down when touched, they are to be reprocessed. After removing the ring, a sealed jar can be tipped without leakage.

QUESTIONS

1. What type of food is processed in the water bath canner and why?

2. What type of food is processed in the pressure canner and why?

3. When canning food, what factors will have an effect on timing the canning process?

4. What is the purpose of 2–3 inches of water in the pressure canner?

5. Does the type of packing of food have an effect on the processing temperature?

II. FOODS TO BE PACKED AND PROCESSED

A. RAW-PACKED TOMATOES (1 PINT JAR)

1. Prepare jar and lid as in Part I.
2. Wash three medium tomatoes. Loosen skins by dipping tomatoes in boiling water for about 30 seconds, then plunge them into cold water. Skins should slip off easily.
3. Cut tomatoes into quarters.
4. Add 1 tablespoon of bottled lemon juice and 1/2 teaspoon of salt to the jar. Place tomatoes into jar.
5. Pour boiling water into jar, leaving 1/2 inch headspace. Remove air bubbles by running spatula between jar and food.
6. Wipe jar rims with a clean towel. Cover with lids and screw on rings firmly, **but not too tightly**. Label lid.
7. Process in a boiling water bath for 45 minutes.
8. Record observations in Table 2.1.

B. HOT-PACKED TOMATOES (1 PINT JAR)

1. Prepare jar and lid as in Part I.
2. Wash three medium tomatoes. Loosen skins by dipping tomatoes in boiling water for about 30 seconds, then plunge them into cold water. Skins should slip off easily.
3. Cut tomatoes into quarters and place them in a saucepan and cover with water. Bring to a boil and boil gently for 5 minutes.
4. Add 1 tablespoon of bottled lemon juice and 1/2 teaspoon salt to the pint jar.
5. Pack hot tomatoes in jars, leaving 1/2 inch headspace.
6. Fill jars to 1/2 inch from the top with hot cooking liquid. Remove air bubbles. Wipe jar rims. Adjust lids and process for 40 minutes in a water bath.
7. Record observations in Table 2.1.

C. APPLES (2 PINT JARS)

1. Prepare jars and lids as in Part I.
2. Wash, pare (peel), and slice six medium apples (**Gala apples work well**) into 1/2 inch slices. Cut slices into thirds. In a large mixing bowl, dissolve 2 tablespoons ascorbic acid powder ("Fruit Fresh") in 4 cups water. Place apple chunks in this solution to prevent darkening.
3. **Hot Pack in Water**
   a. Take half of the apple chunks and put them in a saucepan. Cover with water and bring to a boil. Add 1 tablespoon of bottled lemon juice. Boil for 5 minutes.
   b. Pack apples into one pint jar. Quickly, bring the water where the apples were cooked in to a boil; fill jar with boiling apple water, leaving 1/2 inch headspace. Remove air bubbles by running spatula between jar and food.
   c. Wipe jar rim. Cover with lid and screw on ring firmly, but not too tight. Label lid.
   d. Process in boiling water bath for 25 minutes.
   e. Record observations in Table 2.1.
4. **Raw Pack in Water**
   a. Pack the balance of the cut apples into second pint jar. Quickly, bring 2 cups water plus 1 tablespoon bottled lemon juice to a boil; fill jar with hot water, leaving 1/2 inch headspace. Remove air bubbles by running spatula between jar and food.
   b. Wipe jar rim. Cover with lid and screw on ring firmly but not too tight. Label lid.
   c. Process in boiling water bath for 30 minutes.
   d. Record observations in Table 2.1.

D. PEARS (2 PINT JARS)

1. Prepare jars and lids as directed in Part I.
2. Wash, pare (peel), cut in half, and core six medium pears (such as **Bartlett or Anjou**). Cut pieces into 1/2 inch thick slices. In a large mixing bowl, dissolve 2 tablespoons ascorbic acid powder ("Fruit Fresh") in 4 cups water. Place pear slices in this solution to prevent darkening.
3. Make a syrup by heating 1 cup sugar and 2 cups water until sugar is dissolved. Remove from heat.
4. **Raw Pack in Syrup**
   a. Pack half of the pear slices into one pint jar. Add 1 tablespoon of bottled lemon juice. Bring syrup to a boil, remove from heat, and fill jar with the boiling syrup, leaving 1/2 inch headspace. Remove air bubbles by running spatula between jar and food.
   b. Wipe jar rim. Cover with lid and screw on ring firmly, but not too tight. Label lid.
   c. Process in boiling water bath for 25 minutes.
   d. Record observations in Table 2.1.
5. **Hot Pack in Syrup**
   a. Add remaining pear slices to the remaining syrup. Add 1 tablespoon of bottled lemon juice. Bring to a boil and boil for 5 minutes.
   b. Pack hot pear slices into second pint jar. Quickly, bring syrup to a boil; fill jar with boiling syrup, leaving 1/2 inch headspace. Remove air bubbles by running spatula between jar and food.
   c. Wipe jar rim. Cover with lid and screw on ring firmly, but not too tight. Label lid.
   d. Process in boiling water bath for 20 minutes.
   e. Record observations in Table 2.1.

E. GREEN BEANS (2 PINT JARS)

1. Prepare jars and lids as directed in Part I.
2. Wash 3/4 pound green beans and drain. Cut or break off ends. Cut or break green beans into 1- to 1½-inch pieces.
3. **Raw Pack**
   a. Pack raw beans tightly to 1/2 inch below top of one pint jar. Add 1/4 teaspoon salt (on top of beans).
   b. Cover with boiling water, leaving 1/2 inch headspace. Remove air bubbles by running spatula between jar and food.

    c. Wipe jar rim. Cover with lid and screw on ring firmly, but not too tight. Label lid.

    d. Record observations in Table 2.1.

4. **Hot Pack**

    a. Cover cut beans with boiling water and cook for 5 minutes.

    b. Pack hot beans loosely to 1/2 inch on top of second pint jar. Add 1/4 teaspoon salt.

    c. Bring water that beans were cooked in to a boil. Fill jar with boiling liquid, leaving 1/2 inch headspace. Remove air bubbles by running spatula between jar and food.

    d. Wipe jar rim. Cover with lid and screw on ring firmly, but not too tight. Label lid.

    e. Record observations in Table 2.1.

5. Process both jars in a pressure canner at 10 pounds pressure (240°F) for 30 minutes.

Table 2.1 TABLE FOR EVALUATION OF CANNED PRODUCTS					
Food	Pack	Process	Jar Appearance	*Texture	*Flavor
Tomatoes	Raw				
Tomatoes	Hot				
Apples	Raw				
Apples	Hot				
Pears	Raw				
Pears	Hot				
Green beans	Raw				
Green beans	Hot				

*To be evaluated in a later lab.

F. STRAWBERRY JAM (4 HALF-PINT JARS)

2$\frac{1}{2}$ cups strawberries, crushed

2$\frac{1}{2}$ cups granulated sugar

3 tablespoons + 1/2 teaspoon powdered pectin ("Sure Jel")

1. Prepare jars and lids as directed in Part I.
2. Crush strawberries. Put strawberries and powdered pectin in a large saucepan. **Add 1 cup sugar**.
3. Bring berry mixture to a full boil over **medium heat**, stirring constantly. Immediately stir in the balance of the sugar (**1$\frac{1}{2}$ cups**).
4. Stir and bring to a full rolling boil. Boil hard **1 minute, stirring constantly**.
5. Remove from heat. Skim off foam. Immediately ladle into jars, leaving 1/4 inch headspace.
6. Wipe jar rims. Cover with hot lids and screw on rings firmly, but not too tightly. Label lid.
7. Process in boiling water bath for 10 minutes.
8. Record observations in Table 2.2.

G. FREEZER STRAWBERRY JAM (2 HALF-PINT JARS)

1 cup strawberries, crushed

2 cups sugar

2 tablespoons + 2$\frac{1}{2}$ teaspoons powdered pectin (Sure Jel)

1/4 cup + 2 tablespoons water

1. Prepare jars and lids as directed in Part I.
2. Crush strawberries. Stir sugar into fruit and let stand for 10 minutes.
3. Mix powdered pectin and water in a small saucepan. **Bring to a full boil and boil for 1 minute, stirring constantly**.
4. Immediately stir pectin mixture into fruit and continue to stir for 3 minutes.

5. Ladle mixture into jars, leaving 1/4 inch headspace. Wipe any spills from container. Cover with lid and screw on rings firmly, but not too tight. Label lid.
6. Let stand at room temperature for 24 hours. Store jam in freezer.
7. Record observations in Table 2.2.

H. "LIGHT" GRAPE JELLY (3 HALF-PINT JARS)

2 cups grape juice   2 tablespoons + $2\frac{1}{2}$ teaspoons powdered "light" pectin
1/2 cup water $1\frac{1}{2}$ cups sugar

1. Prepare jars and lids as directed in Part I.
2. Thoroughly mix water and grape juice in a large saucepan.
3. Measure sugar and set aside. Mix **2 tablespoons of measured sugar into powdered "light" pectin**.
4. Stir the pectin–sugar mixture into juice in pan. **Saucepan must be no more than 1/3 full to allow space for full rolling boil**.
5. Bring contents of saucepan to a full boil over **medium heat, stirring constantly**. Stir in remaining sugar.
6. Bring mixture back to a full rolling boil, stirring constantly and boil for 1 minute.
7. Remove from heat. Skim off foam. Pour into hot jars, leaving 1/4 inch headspace.
8. Wipe jar rims. Cover with hot lids and screw rings on firmly, but not too tight. Label lid.
9. Process in boiling water bath for 10 minutes.
10. Record observations in Table 2.2.

Table 2.2 TABLE FOR EVALUATION OF JELLY/JAM			
Product	Clarity	Flavor*	Texture*
Strawberry jam			
Freezer strawberry jam			
"Light" grape jelly			

*To be evaluated in a later lab.

QUESTIONS

1. What is headspace and what functional role does it play in the canning process?

2. What may cause liquid to be lost from jars during canning? Can you open the jar and pour some liquid back after processing?

3. Compare the processing times for raw pack and hot pack food items and account for the differences. Are there any differences in their appearance after processing?

4. What are the main ingredients for making jelly? What causes the pectin to set?

18

5. Under what condition(s) are *Clostridium botulinum* spores destroyed?

6. Was there a difference in the appearance of the fruit packed in water and the fruit packed in sugar? Why?

III. FREEZING OF FRUITS AND VEGETABLES

A. TO SHOW THE EFFECTS OF BLANCHING ON THE QUALITY OF FROZEN VEGETABLES

1. Prepare for freezing 1/4 pound of the following: carrot, broccoli, or cauliflower. Pick over vegetables carefully; discard any rotten or decayed parts; wash and dry thoroughly.
2. Blanch 1/2 of the assigned vegetable:
   a. Place vegetable in basket or strainer.
   b. Plunge basket into boiling water. **Water must be boiling and vegetable must be totally immersed in the water**.
   c. Blanch carrot for 3 minutes; blanch broccoli (flowerets) for 3 minutes; blanch cauliflower (1 inch pieces) for 3 minutes. Then plunge the blanched vegetable in ice water. **The vegetable remains in the ice water the same length of time it was blanched**.
3. Leave the other half of the assigned vegetable unblanched.
4. Place unblanched and blanched vegetable, separately, in small plastic bag; press out any excess air; seal and label. **A vacuum sealer can also be used. Follow manufacturer's instructions for filling and sealing bags**.
5. Freeze at 0°F (−18°C) and hold in frozen storage, preferably for 3 weeks or longer.
6. Cook (**after storage for 3 weeks or more**) the two lots of frozen vegetables until tender in a small amount of boiling salted water.
7. Evaluate the two lots of vegetables for the characteristics listed in Table 2.3.

Table 2.3 TABLE FOR THE EVALUATION OF UNBLANCHED/BLANCHED VEGETABLES					
Vegetable	Treatment	Color	Aroma	Taste	Texture
Carrot	Unblanched				
Carrot	Blanched				
Broccoli	Unblanched				
Broccoli	Blanched				
Cauliflower	Unblanched				
Cauliflower	Blanched				

QUESTIONS

1. Why should vegetables be blanched before they are frozen?

2. Why were the vegetables plunged in ice water after blanching?

3. What effect does blanching have on the nutrient content of the frozen vegetables?

B. TO SHOW THE EFFECTS OF VARIOUS TREATMENTS ON THE QUALITY OF FROZEN FRUIT

1. Wash, pare, and slice six apples (**Red Delicious works best for this exercise**) and divide into six equal portions (one apple for each treatment. WORK WITH EACH APPLE SEPARATELY FOR EACH TREATMENT. DO NOT PEEL ALL THE APPLES AT ONCE. HAVE ALL THE INGREDIENTS READY FOR EACH TREATMENT. TIMING IS IMPORTANT.

2. **Dry Pack**
   a. Leave the first cut apple untreated.
   b. Blanch the second apple for 2 minutes. Cool in ice water for 2 minutes and drain.
   c. Mix 2 tablespoons sugar with the third cut up apple.
   d. Mix 2 tablespoons sugar with 1/8 teaspoon citric acid with the fourth cut up apple.

3. **Syrup Pack (40% Sugar Syrup: 3 cups sugar and 4 cups water. Mix water and sugar together until sugar dissolves).**
   a. To one-half (1/2) cup of the 40% syrup, add 1/8 teaspoon of citric acid and cover the fifth apple.
   b. For the sixth apple cover with 1/2 cup of 40% syrup.

4. **Packaging and Freezing**
   a. Pack into freezer bags the individual treatments. Make sure that all air is expelled from the bag; seal and label. **If available a vacuum sealer can be used for this exercise. Follow manufacturer's instructions.**
   b. Freeze at 0°F and hold in frozen storage, preferably 3 weeks or longer.
   c. Remove from the freezer and thaw, sealed in the refrigerator (4–6 hours) or under cold running water for approximately 30 minutes or longer.
   d. Rank the six samples for color, texture, and flavor in Table 2.4.

Table 2.4 TABLE FOR EVALUATION OF FROZEN APPLES				
Treatment	Pack	Color	Texture	Flavor
Plain	Dry			
Blanched	Dry			
Sugar	Dry			
Sugar + citric acid	Dry			
40% syrup	Syrup			
40% syrup + citric acid	Syrup			

QUESTIONS

1. What causes browning of apples?

2. How can browning be prevented?

3. What treatment should have the best effect on preventing browning and why?

4. How will blanching affect the fruit?

5. Why is it important to have an airtight seal when freezing the fruits and vegetables?

6. How may fluctuations of storage temperature affect the appearance, texture, and nutrient quality of frozen fruits and vegetables?

LABORATORY 3

Starch and Cereal Cookery:
Role of Gelatinization and Gelation

LABORATORY 3
Starch and Cereal Cookery:
Role of Gelatinization and Gelation

Starch is a term used to indicate molecules and the collection of these molecules organized as granules. Starch grains are used as thickening agents in soups and sauces. Cereal cookery is basically starch cookery, as starch makes up the major portion of the cereal grain. The first part of this laboratory exercise is to provide some basis for an understanding of the behavior of starch when used as a thickening agent. The second part stresses the cooking of various cereal grains which were processed under different conditions.

Vocabulary

al dente	enriched	polenta
amylopectin	gelatinization	quick cooking
amylose	gelation	roux
bran	germ	semolina
cornmeal	grits	suspension
converted	instant cereal	viscosity
dextrin	oatmeal	waxy starch
endosperm		

OBJECTIVES

1. To study the effects of moist and dry heats on the cooking properties of starch.
2. To differentiate between gelatinization and gelation.
3. To examine the effects of time, temperature, agitation, acidity, and ingredients on the gelatinization and gelation properties of starch.
4. To recognize and understand that cereal cookery is basically the gelatinization of starch.

PRINCIPLES

1. When starch and water are mixed together, a temporary suspension is formed.
2. When moist heat is applied, the starch granules swell up to a certain point and a colloidal dispersion is formed. This is called *gelatinization*.
3. There is also an increase in viscosity and this is attributed to amylopectin.
4. When the dispersion cools, gelling occurs and this is attributed to the amylose.
5. Heat, time, agitation, acid, and other ingredients will have an effect on gelatinization and gelation.
6. Cereal grains are cooked in water until they become hydrated, tender, and soft in the process of gelatinization.
7. Cooking grains in only the amount of water that will be absorbed when they are fully hydrated permits maximum retention of nutrients and prevents deleterious color changes.
8. The cooking time for cereal products (grains, rice, pasta) depends on the size and characteristics of the grain or pasta.

I. STARCH PRINCIPLES

A. TO SHOW FACTORS WHICH AFFECT THE THICKNESS OF A COOKED STARCH PASTE

As influenced by the kind of starch.
As influenced by sugar, acid, and dextrinization.

1. Use pans of the same size.
2. Use **1 cup water** for each starch listed except **#5** in Table 3.1, and for this use **2/3 cup water and 1/3 cup + 2 tablespoons freshly squeezed lemon juice**.

3. Mix the starch and liquid together. Use a whisk to make sure there are no lumps. Place pan on heat source and cook the mixture over **medium heat, stirring constantly. Bring mixture to a boil and allow it to boil for 1 minute.**
4. Remove from heat. Pour approximately 3/4 of cooked starch paste into a custard cup.
5. Let stand to cool.
6. Unmold onto a small plate and record observations in the table provided.

		Amount of Starch	Firmness	Appearance
Number	Starch	(Tablespoon)	(Cooled Paste)	(Cooled Paste)
1	Flour	2		
2	Cornstarch	1		
3	Cornstarch	2		
4	Modified food starch	2		
5	Cornstarch, water, lemon juice	2		
6	Flour, lightly browned	2		
7	Flour, darkly browned	2		
8	Cornstarch and 1/3 cup sugar	2		

Table 3.1   TABLE FOR THE EVALUATION OF STARCH GELS

QUESTIONS

1. Is there a difference in the thickening and gelling effect between flour and cornstarch? If so, why?

2. How would you recommend the addition of an acid to a starch paste?

3. What effect did dry heat have on flour? Did time and temperature have an effect on gelatinization and viscosity?

4. What effect does sugar have on the gelatinization of starch?

5. Differentiate between gelatinization and gelation. Which component of the starch granule is involved with each process?

B. TO STUDY THE EFFECT OF WATER TEMPERATURE ON STARCH DISPERSION

1. **Cold water**

    a. Stir 1 tablespoon of flour into 8 oz. of cold water in a glass measuring cup. Record your observations.

    _____

    _____

    b. Allow the mixture to stand for 10 minutes. Record your observations.

    _____

    _____

2. **Boiling water**

    a. Bring 1 cup of water to a boil in a 1-quart saucepan. Add 1 tablespoon flour to the boiling water. Stir continuously. Record your observations.

    _____

    _____

QUESTIONS

1. Why does the flour settle to the bottom of the container?

2. What causes the lumps in the heated solution?

3. What is a starch slurry, and how effective is this mixture when it is added to a boiling liquid?

C. TO SHOW THE EFFECTS OF THE PROPORTIONS OF FLOUR TO MILK ON THE CONSISTENCY OF WHITE SAUCE

1. Prepare each of the white sauces in Table 3.2, using **fluid skim milk**.
2. Compare the white sauces for thickness. **Make your observations immediately. If the sauces are allowed to stand for any length of time, they will all thicken and look alike.**

	Table 3.2	BASIC WHITE SAUCE PROPORTIONS			
Type	Flour	Fat	Salt	Liquid	Use
Thin	1 tablespoon	1 tablespoon	1/4 teaspoon	1 cup	Cream soups
Medium	2 tablespoons	2 tablespoons	1/4 teaspoon	1 cup	Creamed dishes, scalloped dishes, gravy
Thick	3 tablespoons	2 tablespoons	1/2 teaspoon	1 cup	Cooked salad dressings, soufflés
Very thick	4 tablespoons	$2\frac{1}{2}$ tablespoons	1/2 teaspoon	1 cup	Croquettes

D. METHOD 1

1. Melt fat over low heat. Add flour and salt and blend.
2. Remove pan from heat; whisk in milk.
3. Return pan to heat; cook over medium heat, stirring constantly until mixture thickens and boils. Boil for 1 minute.
4. Remove from heat.
5. Record observations in Table 3.3.

E. METHOD 2 (LOW-FAT METHOD)

1. Blend the flour and salt with 1/4 cup cold milk. Stir until all lumps have been separated.
2. Add remaining milk. Stir thoroughly.
3. Place the mixture in a saucepan over direct medium heat; stir constantly until the mixture boils; boil and stir for 1 minute.
4. As the mixture boils, add 1/2 of the fat indicated for the type of sauce being prepared. Stir thoroughly until the fat is blended into the sauce.
5. Record observations in Table 3.3.

	Table 3.3 TABLE FOR EVALUATION OF WHITE SAUCE		
Type	Method	Appearance	Texture/Consistency
Thin	1		
Thin	2		
Medium	1		
Medium	2		
Thick	1		
Thick	2		
Very thick	1		
Very thick	2		

CHARACTERISTICS OF HIGH QUALITY WHITE SAUCES

Appearance: White to creamy (dependent on the fat used); opaque.
Consistency: Smooth; even starch gelatinization and distribution.
Thin: Flows freely; like thin cream.
Medium: Fluid, but thick; flows slowly, like heavy cream.
Thick: Thick; hold imprint of spoon.
Very thick: Will not flow; hold cut edge.
Flavor: Very bland; fat used may vary flavor.

QUESTIONS

1. What is the roux method?

2. Why is the sauce boiled for 1 minute?

F. VANILLA PUDDING (CONVENTIONAL METHOD)

1 tablespoon cornstarch   2 egg yolks
2 tablespoons sugar   1/8 teaspoon salt
1 cup milk   1/2 teaspoon vanilla

1. Mix together cornstarch and sugar
2. Add milk, egg yolks, and salt; **whisk well to make sure that the yolk is completely blended with the milk**.
3. Cook over medium heat, stirring constantly.
4. Bring mixture to a boil and boil for 1 minute.
5. Remove from heat and stir in vanilla.
6. Pour into custard cup and cool.
7. Compare to other pudding variations and record observations in Table 3.4.

G. MICROWAVE PUDDING

1. Use ingredients for conventional vanilla pudding.
2. Mix dry ingredients in a quart measuring cup.
3. Gradually stir in milk and egg yolks; continue stirring until thoroughly mixed.
4. Microwave uncovered on high for 3 minutes.
5. Stir well. Microwave on high for 3 minutes more.
6. Remove; stir in vanilla.
7. Pour into serving dish.
8. Compare to other pudding variations and record observations in Table 3.4.

H. INSTANT PUDDING MIX

1. Prepare a vanilla instant pudding mix according to the package directions and compare to the puddings in Table 3.4.

I. COMMERCIALLY CANNED PUDDING

1. Compare a commercially canned pudding to the other puddings and record observations in Table 3.4.

Table 3.4 TABLE FOR EVALUATION OF PUDDING			
Pudding	Appearance	Taste	Consistency
Conventional			
Microwaved			
Instant mix			
Commercially canned			

CHARACTERISTICS OF HIGH-QUALITY PUDDING

Appearance: *Moist and shiny; film will form on top as pudding cools.*
Consistency: *Pudding should be firm but not hard.*
Flavor: *Slightly sweet.*

QUESTIONS

1. What was the purpose of mixing the cornstarch and sugar together in the initial step in preparation of the conventional and microwaved puddings?

2. What is the difference between the puddings?

    a.   Flavor

    b.   Texture

    c.   Ease of preparation

3. How can the skin be prevented from forming on the surface of the pudding?

4. What causes the pudding to set as it cools?

5. What would happen if the pudding was stored in the refrigerator for a long period of time?

II. CEREAL COOKERY

A. TO DEMONSTRATE AND COMPARE THE BEHAVIOR OF STARCH IN CEREAL PRODUCTS

1. Cereals

    a. *Bulgur wheat (sometimes called parboiled wheat)*: Whole wheat that has been cooked, dried, partly de-branned, and cracked into coarse, angular fragments. Originated in the Near East. **Prepare following directions on the package. Record observations in Table 3.5.**

    b. *Farina (granulated wheat endosperm)*: Made from wheat other than durum with the bran and most of the germ removed. It is prepared by grinding and sifting the wheat to a granular form (marketed as Cream of Wheat). **Prepare following directions on the package. Record observations in Table 3.5.**

    c. *Oatmeal (rolled oats)*: Made by rolling the grouts (oats with hull removed) to form flakes. Quick-cooking oats are cut into tiny particles which are then rolled into thin, small flakes. Instant oatmeal has been precooked, and then the water is removed. **Prepare following directions on the package. Record observations in Table 3.5.**

    d. *Hominy grits (corn grits, grits)*: Prepared from either white or yellow corn from which the bran and the germ have been removed. The remaining edible portion is ground and sifted. Grits are coarser than cornmeal. **Prepare following directions on the package. Record observations in Table 3.5.**

    e. *Cornmeal*: Prepared by grinding cleaned white or yellow corn to a fineness specified by federal standards. Cornmeal can be bolted (further decreases size of granule); it may be degerminated (remove germ portion of kernel thus removing the fat); it may be enriched (add specified thiamin, riboflavin, niacin, iron, and folic acid; optional: calcium, vitamin D). **Prepare recipe for cornmeal mush (Recipe B). Record observations in Table 3.5.**

B. CORNMEAL MUSH

1/2 cup cornmeal	1/2 cup cold water
1$\frac{1}{2}$ cups boiling water	1/2 teaspoon salt

1. Bring 1$\frac{1}{2}$ cups water to a boil.
2. Blend the cornmeal, salt, and 1/2 cup cold water together.
3. Remove the boiling water from the heat, spoon or pour the cornmeal mixture into hot water; stir until blended.
4. Return to the heat and bring to a boil; boil for 5 minutes. Stir occasionally.
5. Reduce heat to lowest temperature. Cover; allow to heat for 10 additional minutes, occasionally.

CHARACTERISTICS OF HIGH-QUALITY COOKED CEREALS

Appearance: *Distinct particles, granules, or flakes.*
Consistency: *Thick; somewhat viscous (without gumminess).*
Flavor: *Bland (cooked starch); typical for grain (wheat, corn, oats); well rounded (no raw taste).*
Mouth feel: *Particles remain discrete; soft.*

Table 3.5 TABLE FOR EVALUATION OF CEREALS			
Cereal	Appearance	Consistency	Taste
Oat bran			
Farina, regular			
Farina, instant			
Oatmeal, regular			
Oatmeal, minute			
Oatmeal, instant			
Grits, regular			
Grits, instant			
Cornmeal, white			
Cornmeal, yellow			

C. SPOONBREAD

2 cups milk
1 cup water
2 tablespoons margarine or butter
1/2 cup yellow cornmeal

1/2 cup white cornmeal
1 teaspoon salt
3 large eggs, room temperature

1. Preheat oven to 350°F.
2. Combine milk, water, white and yellow cornmeals, margarine, and salt in a medium-sized saucepan. Cook over medium heat until thickened, stirring constantly. Remove from heat.
3. Crack eggs into a medium-sized bowl. Using a hand mixer, beat the eggs at high speed until thick and lemon colored (about 4–5 minutes).
4. Gradually stir about one-fourth of the hot mixture into the beaten eggs (this is known as **tempering**). Mix thoroughly. Then add the egg mixture to the hot cornmeal mixture.
5. Beat the mixture with the hand mixer at high speed until ingredients are incorporated.
6. Pour mixture into a lightly greased 11/2-quart soufflé dish. Bake in the preheated oven for 35 minutes or until a knife comes out clean.

D. BULGUR PILAF

1 medium onion, chopped
1/2 cup celery, chopped
1/2 red pepper, diced
1/2 small zucchini, diced

3 tablespoons butter or margarine
1 cup bulgur wheat
2 cups chicken, beef, or vegetable broth
1/4 teaspoon salt

1. In a medium-sized skillet melt butter. Sauté celery, onion, red pepper, and zucchini until softened.
2. Add bulgur wheat and mix thoroughly with sautéed vegetables.
3. Add broth and salt.
4. Bring broth to a boil. Cover and reduce heat, and simmer for 15 minutes. Fluff with a fork and let bulgur wheat stand covered for 5 minutes to absorb any excess moisture.

E. RICE

1. Steamed Rice

1/2 cup rice*
1 cup water

1/4 teaspoon salt

*For converted rice follow recipe on the package.

1. Add salt to water and bring to a boil.
2. Stir in rice with a fork.
3. Heat uncovered until water returns to a boil. Then lower heat to simmering; cover pan tightly and cook rice very slowly for 15 minutes. **Do not lift the cover during the cooking period**.
4. Remove cooked rice from the heat, fluff rice with a fork, cover again and allow cooked rice to stand for 5 minutes before serving.
5. Measure the cooked rice to obtain increase in volume: 1/2 cup raw rice yields _____ cups cooked rice.

2. Rice Pilaf

2 tablespoons butter or margarine
1 small onion, finely chopped
1 cup long-grain converted rice

2 cups water*
2 beef, chicken, or vegetable bouillons*
Dash of pepper

*Chicken, beef, or vegetable broth can be used in place of the water and bouillon cubes.

1. Melt butter or margarine in a heavy 2-quart saucepan. Sauté onion in melted butter over low heat until soft and translucent.
2. Add rice and cook for a few minutes until a light brown.
3. Add pepper, water, and bouillon. Bring water to a boil, then lower the heat to a simmer and cover the pan with a tight fitting cover.
4. Cook for 15–20 minutes. Remove from heat; fluff rice with a fork; cover and allow cooked rice to stand for 5 minutes before serving.

3. Curried Rice Pilaf

2 tablespoons vegetable oil
1 large onion, thinly sliced
$1\frac{1}{2}$ teaspoons curry powder
$1\frac{1}{2}$ cups basmati or Texmati or long-grain white rice
1 cinnamon stick

3 cups chicken broth
1 medium red pepper, cored, seeded, thinly sliced
1 small carrot, pared and shredded
1/2 cup whole kernel corn, defrosted

1. Heat oil in large skillet over medium heat. Add onion; sauté until browned, about 7 minutes.
2. Stir in curry powder and rice.
3. Add cinnamon stick and chicken broth.
4. Bring to a boil. Lower heat; cover and simmer for 10 minutes.
5. Add shredded carrot, red pepper, and kernel corn; cover for 5 minutes, simmering.
6. Remove from heat and let stand covered for 10 minutes.

4. Fried Rice

3 cups water
1 teaspoon salt
1½ cups long-grain rice
4 slices turkey bacon
2 tablespoons vegetable oil
2 cloves garlic, minced
1 small onion, minced
3 cups broccoli flowerets, fresh or frozen
2 carrots, peeled and sliced thinly on an angle

5 oz. can sliced water chestnuts
1/4 teaspoon salt
1/8 teaspoon pepper
2 teaspoons minced fresh ginger
1/2 tablespoon dark sesame oil
3 tablespoons low sodium soy sauce
1 tablespoon hoisin sauce
1/2 cup egg substitute or 2 eggs, beaten
4 scallions, sliced thinly on an angle

1. Combine water and salt in a 3-quart saucepan. Cover and bring to a boil. Stir in rice; reduce heat, cover and simmer until rice is tender for 15–20 minutes; fluff with fork; cover and remove from heat; allow to stand for 5 minutes. **Spread cooked rice on a cookie sheet to cool it fast. Usually, fried rice is prepared from previously cooked rice. This suggested method speeds up the cooling process.**
2. Add 1 tablespoon oil to a wok or a wide deep frypan. Add turkey bacon to the pan and cook until bacon is crisp. Remove bacon and place on paper towel to drain. Crumble bacon when cool.
3. Add onion to the wok and stir-fry for 3 minutes; add garlic and stir-fry for 1 minute. Remove onion and garlic and set aside.
4. Add 1 tablespoon vegetable oil to the wok; stir-fry broccoli and carrots for 3–5 minutes. Add cooked onion and garlic to the vegetables. Add salt, pepper, ginger, and water chestnuts. Mix ingredients.
5. Add rice; cook and stir for 3 minutes. Make a well in the center of the rice and add the eggbeaters; with a fork stir the eggs gently until they start to set; then with the spatula stir the rice–vegetable mixture to distribute the cooked egg.
6. Add the sesame oil, soy sauce, and hoisin sauce and distribute evenly. Add reserved crumbled bacon and the sliced scallion as a garnish.

CHARACTERISTICS OF HIGH-QUALITY COOKED RICE

Appearance: *Grains intact; white; translucent.*
Texture: *Grains firm, but tender; fluffy.*
Flavor: *Bland.*

QUESTIONS

1. Why does the amount of water used to cook cereals vary?

2. Describe the nutritional importance of the bran, germ, and the endosperm.

3. Describe how and why starting rice to cook in cold water versus starting it in boiling water affect the texture of the cooked rice.

4. What is meant by the term "pilaf?"

5. How does the amylose content of rice affect its cooking quality?

F. PASTA

1. Baked Spaghetti

1/2 pound spaghetti, cooked until tender	2 (10¾ oz. each) cans condensed tomato soup
1 pound ground round	1 (6 oz.) can tomato paste
1/2 cup chopped onion	2½ cups water
1 large garlic clove, minced	1/4 cup grated Parmesan cheese
1/2 teaspoon basil leaves, crushed	1½ teaspoons salt
1/4 teaspoon thyme leaves, crushed	1/4 teaspoon black pepper
1 tablespoon vegetable oil	1 medium bay leaf

1. Cook spaghetti in a large pot of salted water until al dente. Drain and set aside.
2. In a large saucepan, brown beef in oil; add onion, basil, thyme, and garlic. Cook until onion is translucent.
3. Stir in soup, water, tomato paste, 2 tablespoons of Parmesan cheese, salt, pepper, and bay leaf. Bring to a boil; reduce heat; cover and cook over low heat for 15 minutes. Stir sauce occasionally.
4. Preheat oven to 350°F. Grease a 13 × 9 × 2 inch baking dish; set aside.
5. Combine sauce with the cooked spaghetti. Pour mixture into the prepared baking dish. Sprinkle with additional Parmesan cheese. (**The sauce will be thin, but as the mixture bakes it will thicken, since the pasta will absorb and thicken the sauce.**)
6. Bake for 25 minutes or until mixture is bubbling. Stir before serving. Serve with additional Parmesan cheese.
7. *Variation*: 1/2–1 cup chopped green pepper and/or 3 oz. of sliced mushrooms can also be used in the recipe. Add the vegetables when cooking the beef, onions, and garlic.

2. Chicken Piccata Pasta Toss

3/4 pound bowtie pasta (farfalle), cooked according to the package directions	2 tablespoon butter, divided
2 tablespoons olive oil, divided	2 tablespoons flour
1½ pounds chicken tenders, cut into 1 inch pieces	1/2 cup white wine
Salt and pepper to taste	Juice of 1 lemon
4 cloves of garlic, minced	1 cup chicken broth
2 medium shallots, chopped	3 tablespoons chopped parsley

1. Cook pasta according to the package directions. This should be done just before the chicken and sauce is cooked and ready to be combined with the pasta.
2. Heat a deep nonstick skillet over medium heat. Add 1 tablespoon olive oil; heat oil and add chicken to the pan.
3. Season with salt and pepper; cook until lightly golden brown all over, about 5–6 minutes (let chicken cook for 3 minutes before turning). Remove chicken from the pan and set it in a serving dish while you complete the sauce.
4. Return the skillet to the heat; reduce the heat to medium. To the skillet add 1 tablespoon olive oil and 1 tablespoon butter. Add shallots and garlic and sauté for 2 minutes (make sure that the garlic does not over brown).
5. Stir in flour and cook for 2 minutes. Whisk in wine, lemon juice, and chicken broth. Bring to a boil and boil for 1 minute.

6. Stir in remaining 1 tablespoon butter and 1½ tablespoons chopped parsley. Return cooked chicken into the pan and heat until bubbly.
7. Toss sauce with hot cooked pasta. Sprinkle with reserved chopped parsley and Parmesan cheese, if desired.

3. Marinara Sauce

2 tablespoons olive oil
2–3 garlic cloves, finely chopped
1/3 cup chopped parsley
1/2 teaspoon salt
1 can (28 oz.) Italian plum tomatoes, broken up

1/2 teaspoon oregano
1/2 teaspoon basil
Dash pepper
8 oz. spaghetti

1. Sauté olive oil, garlic, and parsley.
2. Add tomatoes, oregano, basil, salt, and pepper. Mash tomatoes with a fork.
3. Simmer uncovered for about 30 minutes or until thickened.
4. Serve over hot cooked spaghetti.

4. Noodles Alfredo

8 oz. fettuccine or linguine
2 tablespoons butter or margarine
1 clove garlic, minced
1 tablespoon flour
1/4 teaspoon salt
1½ cups skim milk

2 tablespoons fat-reduced cream cheese
1 cup grated Parmesan cheese
Dash freshly grated nutmeg
2 teaspoons chopped fresh flatleaf parsley
Cracked black pepper

1. Cook pasta according to the package directions; drain, reserving 2 tablespoons cooking water. Return water and pasta to pot.
2. Meanwhile, in a medium saucepan, melt butter on medium heat. Add garlic; cook for 1 minute, stirring frequently. Stir in flour and salt; gradually add milk, stirring with a whisk. Cook for 6 minutes over medium heat or until mixture thickens and boils, stirring constantly.
3. Add 1/2 cup Parmesan cheese, the cream cheese, and dash of nutmeg; stirring with a whisk until the cheeses melt.
4. Toss sauce with hot pasta. Sprinkle with remaining Parmesan cheese. Garnish with parsley and pepper.

5. Neopolitan Casserole

3/4 pound ground turkey
1/2 cup onion, finely chopped
1/2 cup green pepper, chopped
1 clove garlic, crushed
1/2 teaspoon dried basil
1/2 teaspoon oregano
1/2 teaspoon fennel seeds
1/8 teaspoon red pepper flakes

1 tablespoon sugar
3/4 teaspoon salt
1 can (1 pound 14 oz.) Italian-style tomatoes, undrained
1/2 pound fresh spinach or 1 (10 oz.)
    packaged frozen chopped spinach
    (defrosted and drained well)
1/4 pound medium shell macaroni, cooked
1/2 cup low-fat mozzarella cheese, grated

1. Sauté ground turkey, onion, green pepper, garlic, basil, oregano, fennel seeds, and red pepper flakes; stir frequently until turkey is brown and vegetables are tender—about 15 minutes.
2. Add sugar, salt, and tomatoes. Mash tomatoes with a wooden spoon. Bring to boiling. Reduce heat; simmer uncovered and stir occasionally until thickened—about 15–20 minutes.
3. Wash spinach thoroughly and remove stems. Place in a kettle with some boiling water. Cook covered until wilted about 2–4 minutes. Drain well; reserve.
4. Preheat oven to 350°F.

5. In a large pot cook shell macaroni in salted boiling water until al dente; drain well. In large kettle or bowl combine sauce, spinach, and shell macaroni. Turn mixture out into a greased 1½-quart casserole dish.
6. Sprinkle with grated mozzarella cheese. Bake uncovered for 30 minutes or until bubbly and lightly browned.

6. Baked Ziti with Vegetables

2 tablespoons olive oil
1 medium-sized sweet green pepper, cored, seeded, and diced
1 medium-sized red pepper, cored, seeded, and diced
1 medium-sized yellow pepper, cored, seeded, and diced
2 large cloves garlic, chopped
2 large onions, chopped
1/4 pound mushrooms, chopped
1 can (16 oz.) whole tomatoes, undrained

1 can (8 oz.) tomato sauce
1 teaspoon basil, crumbled
1/2 teaspoon salt
1/4 teaspoon black pepper
1/4 teaspoon leaf oregano
1/2 pound fresh spinach, cleaned and stemmed
8 oz. ziti, cooked according to the package directions
8 oz. part-skim ricotta cheese
1/4 cup grated Parmesan cheese

1. Heat 1 tablespoon olive oil in a large skillet over medium heat. Add green pepper, red pepper, yellow pepper; sauté until barely tender, about 5 minutes. Remove with a slotted spoon and set aside.
2. Heat remaining tablespoon olive oil in skillet; add onion; sauté until softened for 4–5 minutes. Add garlic and mushrooms; sauté for 2 minutes.
3. Break up tomatoes with a fork and add with liquid to the skillet along with the tomato sauce, salt, basil, black pepper, and oregano. Bring to a boil. Lower heat; simmer uncovered until slightly thickened, about 15–20 minutes. Add spinach; cook; stirring until spinach wilts.
4. Preheat oven to 350°F. Spray a 2½-quart casserole with a nonstick vegetable spray.
5. Combine cooked ziti, tomato mixture, ricotta cheese, and 2/3 of the cooked peppers in a large bowl. Spoon mixture into prepared casserole. Sprinkle with Parmesan cheese. Bake for 25 minutes. Sprinkle with remaining pepper mixture.

CHARACTERISTICS OF HIGH-QUALITY PASTA

Appearance:	*Distinct strands or pieces.*
Tenderness:	*Tender; little resistance to bite.*
Flavor:	*Bland; noodles may have a slight egg flavor.*

GENERAL QUESTIONS

1. What is the difference between

    a. grits and cornmeal?

    b. regular, instant, and quick-cooking oatmeal?

    c. long-grain rice and converted rice?

2. Why must pasta be cooked in a large amount of water?

3. What is the difference in the ingredients between various pasta products?

4. What is enrichment and why are cereal products enriched?

5. To what degree should pasta be cooked when it is to be used in a casserole recipe?

6. What is the term to describe the change that a cereal product undergoes when subjected to moist heat?

7. What is the difference between a whole grain product and a refined grain product?

LABORATORY 4

Quick and Yeast Breads:
Role of Manipulation and Gluten Formation in Doughs

Quick and Yeast Breads: Role of Manipulation and Gluten Formation in Doughs

Bread has an integral role in the diet. There are quick breads (biscuits, muffins, etc.) which require very little manipulation, thereby having very little gluten formation, and are leavened by baking powder and/or baking soda. Yeast bread requires a longer mixing time, greater gluten formation, and requires yeast as the leavening agent. This laboratory exercise will introduce the student to the making of these various breads. The exercise will also emphasize the various leavening agents used, as well as the selection of various flours and their role in the structural formation in the breads.

VOCABULARY

all-purpose flour
baking powder
baking soda
biscuit method
bread flour
cut in

fermentation
gluten
glutenin
gliadin
knead

muffin method
oven-spring
proofing
soft wheat flour
yeast

OBJECTIVES

1. To study the effect of manipulation on gluten development in quick breads.
2. To identify the gluten properties of various wheat products.
3. To learn how different leavening agents affect volume and structural development in both quick and yeast breads.

PRINCIPLES

1. Gluten, which gives structure to any baked product, is made up of gliadin and glutenin.
2. The strength of gluten formation is dependent on
   a. the amount of protein in the wheat flour,
   b. the amount of manipulation.
3. Muffins and biscuits require very little mixing for gluten development.
4. Muffin ingredients (liquid added to dry ingredients) are stirred for about 15 strokes (batter will look lumpy).
5. Biscuit dough is kneaded 10 strokes and the dough is rolled 1/2–3/4 inch thickness for cutting biscuits.
6. Yeast dough is kneaded for 8 minutes to
   a. develop gluten,
   b. distribute the ingredients.
7. One important ingredient in yeast dough is yeast.
8. Yeast is allowed to ferment and this causes the dough to rise.
9. The temperature of the water should be between 105 and 115°F to ensure complete activation of the yeast.

I. TO EVALUATE FACTORS WHICH AFFECT THE QUALITY OF MUFFINS

A. MUFFINS (BASIC RECIPE)

2 cups all-purpose flour
1 teaspoon salt
1 tablespoon double-acting baking powder
2 tablespoons granulated sugar

2 tablespoons vegetable oil
1 egg
1 cup milk

1. Preheat oven to 400°F.
2. Sift together flour, salt, baking powder, and sugar in a medium-sized mixing bowl.
3. Blend together thoroughly egg, oil, and milk in a 2-cup liquid measuring cup.
4. Pour liquid ingredients into the dry ingredients (**muffin method**) and mix ingredients together until dry ingredients are moistened. Batter will be lumpy.
5. Fill greased muffin cups 2/3 full. Place pan in preheated oven and bake for 20–25 minutes.
6. **Variations in Manipulation** (Fig. 4.1)
    a. In Step 3, stir batter **7 strokes** and then spoon batter into two greased muffin cups.
    b. Stir **4 more strokes** and spoon the batter into two more greased muffin cups.
    c. Stir an **additional 4 strokes** and spoon the batter into two more greased muffin cups.
    d. Stir **5 more strokes** or until smooth and shiny. Spoon the batter into last two greased muffin cups.
    e. Record observations in Table 4.1.
7. **Variation using maximum amount of fat and sugar**
    a. Using mixing variations as above.
    b. Substitute in basic recipe.
    c. Use 4 tablespoons granulated sugar and 4 tablespoons vegetable oil.
8. Record all observations in Table 4.1.

FIG. 4.1: From left to right, undermixed, optimum, and overmixed muffin. Far right—cross section of an overmixed muffin shows tunneling that occurs during baking.

Table 4.1 TABLE FOR THE EVALUATION OF MUFFINS				
Variation	7 Strokes	11 Strokes	15 Strokes	20 Strokes
Regular recipe				
Extra fat and sugar				

CHARACTERISTICS OF HIGH-QUALITY MUFFINS

Color:	Golden brown exterior.
External appearance:	Slightly rounded, pebbly tops.
Texture:	Tender and light.
Structure:	Even-textured with medium, rounded holes; slightly moist.

QUESTIONS

1. Describe the method in making muffins.

2. How are muffins leavened?

3. What ingredients are contained in double-acting baking powder? Give the chemical reactions that are involved in the activation of the baking powder.

4. How does under and over manipulation of the muffin batter affect quality of the finished product?

5. What affect did the extra fat and sugar have on the baked quality of the muffins? Why?

6. Name other quick bread products that are prepared by the muffin method.

II. TO EVALUATE FACTORS WHICH AFFECT THE QUALITY OF BISCUITS

A. BASIC BISCUIT RECIPE

2 cups all-purpose flour	5 tablespoons solid fat
3/4 teaspoon salt	2/3–3/4 cup milk (approximate)
1 tablespoon double-acting baking powder	

1. Preheat oven to 425°F.
2. Sift flour, salt, and baking powder together in a medium-sized mixing bowl.
3. Cut in fat (using a pastry blender or two knives) until it resembles coarse cornmeal (**biscuit method**). Add milk and with a fork stir ingredients until they come together.
4. Turn dough onto a lightly floured surface and knead 8–10 times until the dough is smooth.
5. Roll dough out to 1/2–3/4 inch thickness.
6. Cut with a floured biscuit cutter (2–2$\frac{1}{2}$ inches in diameter). Cut straight down into dough and lift straight out. Use even pressure in cutting on the dough to get more evenly shaped biscuits. Do not twist.
7. Bake at 425°F for 10–12 minutes or until lightly browned.
8. Record observations in Table 4.2.

Mixing Variation

1. Follow Basic Biscuit Recipe (above recipe), **Steps 1–3.**
2. Turn dough onto a lightly floured surface, but only **knead 5–6 strokes**.
3. Roll dough out in a rectangle to 1/2–3/4 inch thickness. Fold dough into thirds (**as you would fold a letter**).
4. Turn dough 90° and roll out one more time to 1/2–3/4 inch thickness.
5. Cut and bake biscuits following directives in **Steps 6 and 7** as in the above.

B. BUTTERMILK BISCUITS

1½ cups all-purpose flour	1/2 teaspoon salt
1 teaspoon double-acting baking powder	1/4 cup solid shortening
1/2 teaspoon baking soda	2/3 cup buttermilk (approximately)

1. Preheat oven to 425°F.
2. Sift together flour, baking powder, baking soda, and salt into a medium-sized mixing bowl.
3. Cut in shortening with a pastry blender until it resembles coarse cornmeal.
4. Add milk and stir with a fork until it resembles a soft dough.
5. Turn the dough onto a lightly floured surface and knead about 10 times.
6. Roll out to 1/2–3/4 inch thickness.
7. Cut with a floured cutter. Use even pressure when cutting down on the dough to get evenly shaped biscuits. Do not twist.
8. Place biscuits on an ungreased baking sheet. Bake for 10–12 minutes.
9. Record observations in Table 4.2.

Mixing Variation

1. Follow the above recipe, Buttermilk Biscuits, Steps 1–3.
2. Then follow the directions as in the above (**Basic Biscuit Recipe Mixing Variation**), Steps 2–5.

C. LOW-FAT BUTTERMILK BISCUITS

(Reprinted by permission of Southern Progress Corporation, *The New Complete Cooking Light Cookbook*, 2006, p. 73.)

2 cups all-purpose flour	1/4 teaspoon salt
2 teaspoons baking powder	3 tablespoons + 1 teaspoon chilled butter or margarine
1/4 teaspoon baking soda	3/4 cup low-fat buttermilk

1. Preheat oven to 450°F.
2. Sift together flour, baking powder, baking soda, and salt into a medium-sized mixing bowl.
3. Cut in chilled butter with a pastry blender until mixture resembles coarse cornmeal.
4. Add buttermilk and stir with a fork until ingredients are moistened.
5. Turn dough out onto a floured surface; knead 4–5 times or until dough is just smooth.
6. Roll out dough 1/2–3/4 inch thickness. Cut dough with a floured biscuit cutter.
7. Place cut biscuit dough out onto an ungreased baking sheet.
8. Bake biscuits for 12 minutes or until lightly browned.
9. Record observations in Table 4.2.

Mixing Variation

1. Follow the recipe for Low-Fat Buttermilk Biscuits, Steps 1–3.
2. Then using the **Mixing Variation for Basic Biscuit Recipe, follow Steps 2–5 for mixing and forming of biscuits**.
3. Use baking temperature as directed in the recipe for Low-Fat Buttermilk Biscuits.

D. WHIPPING CREAM BISCUITS

2 cups self-rising all-purpose flour or soft wheat flour for biscuits	1 cup heavy or whipping cream

1. Preheat oven to 450°F.
2. Add flour to a medium-sized mixing bowl. Add whipping cream.
3. Stir ingredients with a fork until a soft dough is formed.
4. Turn dough onto a lightly floured surface and knead about 10 times, until smooth.
5. Roll dough to 1/2–3/4 inch thickness. Cut with a floured biscuit cutter. Use even pressure when cutting down on the dough to get an evenly shaped biscuit. Do not twist.
6. Place cut biscuit dough onto an ungreased baking sheet.
7. Bake in preheated oven for 10–12 minutes.
8. Record observations in Table 4.2.

Table 4.2 TABLE FOR THE EVALUATION OF BISCUITS			
Biscuit	Flakiness	Tenderness	Flavor
Basic			
Buttermilk			
Low-fat			
Whipping cream			

CHARACTERISTICS OF HIGH-QUALITY BISCUITS

Color: *Light and golden.*
Volume: *High and fairly smooth, level tops (Fig. 4.2).*
Texture: *Tender and light.*
Structure: *Flaky and slightly moist.*

FIG. 4.2: A properly mixed biscuit is high, tender, flaky, and contains the traditional crack which allows the biscuit to easily separate.

QUESTIONS

1. What is the method used for mixing biscuits and how does it differ from making muffins?

2. What would happen if the biscuits were over-kneaded?

3. Why was baking soda used in the buttermilk biscuits?

4. Give the reaction that baking soda undergoes in the batter or dough that it is contained in.

   a. How does the reaction of double-acting baking powder differ from the reaction of baking soda?

5. What is another quick bread product that is prepared by the biscuit method?

III. TO IDENTIFY THE GLUTEN-FORMING PROPERTIES OF VARIOUS WHEAT FLOURS

A. GLUTEN BALLS

(Note: Instructor should list type of flours which illustrate various degrees of gluten development; such as, all-purpose flour, bread flour, cake flour, and whole wheat flour)

1. Measure 1 cup flour of each type into a medium-sized mixing bowl. Add just enough water (about 1/4 cup) to form a stiff dough. Rub the wetted flour around the sides of the bowl to incorporate any fragments into the dough. Knead the ball of dough for about 15 minutes to develop the gluten.
2. Fill the bowl with cool water and knead the dough under the water to wash out the starch components. Change the water as needed, using care to retain the gluten. (Or, carefully wash under running water. Place a strainer under gluten ball to collect any pieces. Add back to the ball.)
3. **Washing is completed when the water squeezed from the mass is clear. Place gluten ball(s) onto an ungreased baking sheet**.
4. Bake at 400°F for 15 minutes; reduce the heat to 300°F and continue baking for 40 minutes. Record observations in Table 4.3.

Table 4.3 TABLE FOR EVALUATION OF GLUTEN BALLS	
Type of Flour	Observations

43

QUESTIONS

1. What is the relationship of the size of the gluten ball to the type of flour and amount of protein?

2. What does the gluten ball reveal as to the flour's baking quality?

IV. TO STUDY FACTORS WHICH AFFECT THE QUALITY OF YEAST BREAD

A. BASIC YEAST DOUGH

1/2 cup warm water (105–115°F, or 40°C)   2 cups all-purpose flour*
1 tablespoon granulated sugar   1/2 teaspoon salt
2 tablespoons dried milk solids   2 tablespoons shortening
1 package quick rising yeast   1 egg

1. Preheat the oven to 400°F.
2. Measure the water and place it in a 2-quart bowl. Add the yeast. Stir until blended. Add the sugar and the dried milk solids. Stir until blended. Allow this yeast mixture to stand while measuring the remaining ingredients (5–10 minutes). The yeast activity is initiated.
3. Add the egg and 1 cup of the flour to the yeast mixture. Beat until the batter is smooth (about 100 strokes).
4. Add the salt, shortening, and half of the remaining flour (1/2 cup); stir until well blended.
5. If the dough is still too sticky to turn out onto a lightly floured surface, add the remaining portion of flour and stir into the dough. If the dough is not sticky, use the flour that was not put into the dough to lightly flour (1/2 cup) the work surface.
6. Knead the dough until it is smooth and satiny (about 8–10 minutes).
7. Place the dough in a lightly greased bowl; turn the dough over in the bowl to evenly coat the dough with the fat. Cover the bowl with plastic wrap, and then place the bowl over another bowl filled partly with warm water. Allow dough to rise at least 20 minutes—to double in bulk—and then test if the dough is fully fermented when two fingers leave an indentation in the dough.
8. Lightly knead the dough for 1 minute to evenly distribute the gas cells.
9. Weigh out 75 g of dough. Shape into a bread loaf and place in a small greased bread pan ($4\frac{1}{2} \times 2\frac{1}{2}$ inches). Shape the remaining dough into crescent rolls, clover leaf rolls, or any other shapes that you desire (Fig. 4.3).
10. Allow the loaf of bread and shaped rolls to rise until double in bulk. This may take 20 minutes or more depending on the warmth of the room. An indentation made into the dough should remain as an indicator when the dough has risen sufficiently.
11. Bake bread loaf and rolls for 20–25 minutes.
12. Record observations in Table 4.4.

*Instructor can select various flours for this exercise: bread, cake. When using whole wheat flour, use $1\frac{1}{2}$ cups all-purpose flour and 1/2 cup whole wheat flour

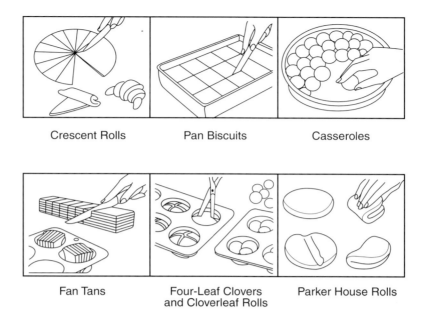

FIG. 4.3: Preparation of various dinner rolls.

Table 4.4 TABLE FOR EVALUATION OF YEAST BREAD			
Type of Flour	Height	Texture	Flavor

CHARACTERISTICS OF HIGH-QUALITY YEAST BREADS

Volume: *High, well-shaped loaf.*
Color: *Uniformly golden brown.*
Texture: *Even, no large air holes.*
Crumb: *Moist and silky, with an elastic quality.*

B. BAGELS

1 package rapid (quick) rising yeast
1 cup warm water (105–115°F)
1 tablespoon granulated sugar, divided
2 teaspoons salt
3 cups all-purpose flour

2 quarts water
1/4 cup granulated sugar
1 teaspoon baking soda
Cornmeal
1 egg + 1 tablespoon water

1. Dissolve yeast and $1\frac{1}{2}$ teaspoons granulated sugar in 1 cup water; allow to rest for 5 minutes (mixture starts to bubble giving a sign that the yeast is alive and fermentation has started).
2. In a large mixing bowl, add $1\frac{1}{2}$ teaspoons granulated sugar, salt, and $1\frac{1}{2}$ cups of all-purpose flour; add yeast mixture; stir until flour is moistened and a thick batter is formed. Add the balance of the flour ($1\frac{1}{2}$ cups).
3. Turn dough out onto a floured surface. Start kneading the dough, and knead for 7 minutes. Keep adding flour to the work surface as needed. **The dough must not be sticky.**
4. Place kneaded dough into a greased bowl; turn the dough around the bowl to grease evenly. Cover the bowl with plastic wrap and place the bowl over another bowl of warm water. Allow the dough to rise for 20 minutes. The dough is ready when an imprint with your finger remains in the dough. **The dough will not rise much.Punch down dough.**
5. Divide dough into 6–8 even balls; roll the individual balls between your hands until smooth; push your thumb through the ball and stretch the dough to form a bagel. Place the formed bagel on a greased plate. **Divide bagels evenly among the plates (3 or 4 on one; 3 or 4 on another).**
6. Let rise **uncovered** for 20 minutes. Start timing when the first formed bagel is placed on the plate.
7. Start heating your water in a large pot. The pot should be filled 2/3 with water. Also, start preheating the oven to 425°F.
8. Just before the bagels are added to the boiling water, add the 1/4 cup granulated sugar and the 1 teaspoon baking soda. The water temperature will drop. Wait for the water to come back to a full boil and then add the bagels, 3 or 4 at a time. **It is imperative that the water is boiling when the formed bagels are added and the water must be vigorously boiling while the bagels are in the water.**
9. **Boil the bagels for 2 minutes on one side.** Turn them over and **boil for $1\frac{1}{2}$ cups minutes** on the other side.
10. Drain boiled bagels on cooling rack for 1 minute.
11. Place bagel on lightly greased baking sheet sprinkled with cornmeal. Brush bagel with egg wash (1 whole egg, whisked with 1 tablespoon water). If desired, sprinkle with topping (see below). Bake in preheated 425°F oven for 25–30 minutes.
12. Remove bagels from baking sheet onto cooling racks.
13. **Variation.** Desired toppings: Sprinkle poppy seeds, sesame seeds directly onto the bagel after it has been glazed before going into the oven. Onion topping: 1 tablespoon dried minced onions + 1 tablespoon water. Allow the onions to hydrate for 5 minutes before being placed on unbaked glazed bagel.

C. ROLLS THAT GRANDMA USE TO MAKE

2 cups skim milk	2 packages rapid rise yeast
1/4 cup butter or margarine, room temperature	1/4 cup lukewarm water (105–115°F)
1/4 cup sugar	7 cups all-purpose flour
2 teaspoons salt	2 eggs

1. Scald milk in saucepan. Pour over butter, sugar, and salt in large mixing bowl. Cool to lukewarm.
2. Sprinkle yeast over lukewarm water; stir to dissolve.
3. Add yeast mixture, 3 cups flour and eggs to milk mixture. Beat mixture with a hand mixer until smooth, about 2 minutes.
4. Gradually stir in enough flour to make a soft dough. Turn dough out onto a floured surface. Knead until smooth and satiny, about 5 minutes.
5. Place dough in a greased bowl, turning over once to grease the top. Cover and let rise in a warm place until doubled, about 30 minutes. Imprint remains when poked with a finger.
6. Punch down dough. Divide dough into thirds. Allow it to rest for 5 minutes.
7. Divide each third into 12 pieces. Shape each piece into a ball. Place 12 balls, equally spaced, in three greased 9-inch round cake pans. Cover and let rise in a warm place until doubled for 25–30 minutes.
8. While rolls are rising, preheat oven to 400°F. When rolls are ready, place in the oven and bake for 15 minutes or until golden brown. Cool rolls in pan for 10 minutes and then remove from pan and cool rolls on racks.

QUESTIONS

1. At what temperature should yeast be dissolved? What would happen if the temperature is too hot or not hot enough?

2. Why should milk be scalded when used in a yeast dough recipe?

3. What functional role does sugar play in yeast dough? What would happen if too much sugar was added to the recipe?

4. What is the role of salt in a yeast dough recipe?

5. What are the three commercial sources of yeast? How do they differ?

6. What is the fermentation formula that the yeast produces in the dough? What are the food sources that are utilized by the yeast for fermentation?

7. What are the reasons for kneading yeast dough?

8. What is meant by proofing the yeast dough? What happens if yeast dough is under proofed? Over proofed?

9. What would be the difference in bread that is made with hard wheat flour and soft wheat flour?

10. How does bread stale? What is added to commercially prepared bread to prevent molding?

11. What is enriched flour and how does it benefit the nutrient content of yeast bread?

12. How much whole wheat flour can be substituted for regular wheat in a bread recipe and why?

LABORATORY 5

Shortened- and Foam-Style Cakes

LABORATORY 5
Shortened- and Foam-Style Cakes

Cake making is an art which requires exact measuring, mixing, and baking temperature. Shortened cakes contain fat, and the type of fat that is selected will affect flavor, texture, and volume. Foam-style cakes consist of angel food and sponge cakes while chiffon cake is included in some instances. Foam-style cake does not contain fat except chiffon (contains oil). All these styles of cakes contain a high ratio of granulated sugar which will affect volume and texture. The student will learn the proper techniques in cake making while learning the proper selection of ingredients and their functional roles.

VOCABULARY

angel food cake
chiffon cake
creaming
dry stage (egg whites)

emulsifier
fold
Maillard browning
one-bowl method

plastic fat
soft peak stage (egg whites)
sponge cake
stiff peak stage (egg whites)

OBJECTIVES

1. To learn the functional role of ingredients in shortened- and foam-style cakes.
2. To learn the proper mixing and folding techniques when preparing shortened- and foam-style cakes.
3. To learn the four stages of beating egg white foams for incorporation into foam-style cakes.
4. To demonstrate the effect of different pH levels in chocolate cake batters.

PRINCIPLES

1. Using the proper ingredients, measurements, and mixing techniques will ensure a properly baked cake.
2. Sugar and fat contribute tenderness and volume to a cake.
3. Plastic-style fats or room temperature fats (butter or margarine) incorporate air into a shortened cake batter which will contribute volume to a baked cake.
4. Creaming the fat and sugar will produce a cake that is high in volume and fine in texture.
5. One-bowl method cakes require a plastic fat (or a room temperature fat, such as butter or margarine) for complete incorporation into the batter.
6. Creaming method cakes and one-bowl method cakes are leavened by air, steam, and CO_2.
7. Foam-style cakes constitute:
   a. Angel food cakes which are made with egg whites.
   b. Sponge cakes which are made with egg yolks and egg whites, or the foam of the beaten whole egg.
   c. Chiffon cakes which are a cross between a foam-style cake and a shortened-style cake (contains oil).
8. The more egg whites are beaten, the more air is incorporated and the larger the volume of foam obtained (whites must be at room temperature for efficient incorporation of air). Egg whites go through four stages during the beating process:
   a. *Foamy Stage*
      i. Egg white is lightly whipped and is frothy and fluid.
      ii. Foam does not hold a peak. At this time add the salt, cream of tartar, and flavoring if specified in the recipe.
   b. *Soft Peak Stage*
      i. Foam appears white, moist, and shiny.
      ii. Foam has small bubbles and flows in the bowl; peaks are formed but the tips of the peaks fold over. During the development of this stage, add sugar one tablespoon at a time if specified in the recipe (**meringue method**).
   c. *Stiff Peak Stage*
      i. Foam does not flow in the bowl, but is still shiny.
      ii. When the beaters are removed, the peaks that are formed remain upright.
      iii. A cut through the foam (no sugar used) leaves a clean path. **It is not advisable to whip egg whites beyond this stage.**

d. *Dry Foam Stage*
    i. Egg whites achieve maximum volume and appear dry and curdled.
    ii. When beaters are removed, foam breaks instead of forming peaks.
    iii. Foam will collapse readily and is lumpy when added to other mixtures.

9. Baking pans for shortened cakes are greased and floured, while for foam-style cakes the tube pan is ungreased.
10. Shortened-style cakes are tested for doneness by any of the following:
    a. a toothpick stuck in the center of the baked cake comes out clean.
    b. the baked cake just starts to pull away from the sides of the pan.
    c. the baked cake springs back when lightly pressed with the finger.
11. Traditional foam-style cakes, such as sponge, angel food, and chiffon, are baked in an ungreased tube pan and are done when they spring back when touched with the finger. A true angel food cake and sponge cake are leavened by **air and steam**.
12. Shortened-style cakes are cooled in the pans for 10 minutes before being removed and placed on a wire rack for complete cooling (at least 2 hours before frosting).
13. Foam-style cakes are cooled upside down in the ungreased tube pan for 2 hours before removal.
14. The pH of chocolate cake batter will have an effect on the color of the baked cake: pH = 7: red devil's food cake; pH < 7: light brown chocolate cake; pH > 7: dark chocolate cake.

HIGH-ALTITUDE ADJUSTMENTS

When baked above 3000 feet, cakes will not rise properly. Use this chart (Table 5.1) as a guide when baking cakes at high altitudes. In addition, when baking a cake above 3000 feet, increase the baking temperature by 25°F.

Table 5.1 HIGH-ALTITUDE ADJUSTMENTS				
Ingredient	3000 feet	5000 feet	7000 feet	10,000 feet
Sugar, for each cup, decrease	1–3 teaspoons	1–2 tablespoons	1$\frac{1}{2}$–3 tablespoons	2–3$\frac{1}{2}$ tablespoons
Liquid, for each cup, add	1–2 tablespoons	2–4 tablespoons	3–4 tablespoons	3–4 tablespoons
Baking powder, for each teaspoon, decrease	1/8 teaspoon	1/8–1/4 teaspoon	1/4 teaspoon	1/4–1/2 teaspoon

I. SHORTENED-STYLE CAKES

To evaluate cakes prepared by the creaming method and the one-bowl method.

A. CREAMING METHOD (CONVENTIONAL METHOD)

1. Yellow Cake

1 cup cake flour
1 teaspoon double-acting baking powder
1/4 teaspoon salt
1/4 cup butter, margarine, or shortening**

2/3 cup granulated sugar
1/2 teaspoon vanilla extract
1 large egg, room temperature
1/3 cup milk

1. Preheat oven to 350°F. Set oven rack to the middle position in the oven.
2. Cut waxed paper to fit the bottom of an 8-inch round layer pan.

3. Grease only the bottom of the cake pan; insert the waxed paper; grease the waxed paper. Set the prepared pan aside. Sift together on a large piece of waxed paper the flour, salt, and baking powder; set aside.
4. Place the butter, margarine, or shortening into a medium-sized bowl. Set the mixer at medium speed and beat the fat until soft and creamy; about 2 minutes. Gradually add sugar to the beaten fat. Total creaming should take 5–7 minutes. This is a critical step.
5. Add the egg to the fat–sugar mixture, and beat until the mass is light and fluffy, about 1–2 minutes; constantly use the rubber spatula to make sure all ingredients are combined and batter is off the walls of the mixing bowl.
6. Add approximately 1/3 of the flour mixture to the egg–fat–sugar mixture. Mix at slow speed initially and with rubber spatula to guide the ingredients into the beaters. Increase speed and mix for 1 minute. Add 1/2 of the milk and mix for another 1 minute; scrapping the bowl constantly.
7. Add second portion of the flour mixture and balance of the milk as stipulated in Step 6.
8. Add the last portion of the flour mixture. Incorporate ingredients carefully, and then increase speed and beat for 1 minute, scrapping the bowl constantly.
9. Push all the batter into the prepared cake pan; level the batter evenly with rubber spatula. Bake for approximately 25 minutes.
10. Do not open oven door until the maximum baking time has elapsed, and then test cake for doneness.
11. Remove cake from oven, and allow the cake to cool for 10 minutes. Then remove cake from pan and place on wire rack to cool thoroughly.

** **Each unit should be assigned a specific fat**. *Butter and margarine should be removed from the refrigerator at least 1 hour ahead to be at room temperature*. **When the cakes are ready, each student should examine and sample the baked cake, and record the results in Table 5.2.**

Table 5.2 TABLE FOR EVALUATION OF SHORTENED-STYLE CAKE MADE BY THE CREAMING METHOD				
Fat	Crust Color	Grain	Moistness	Flavor
Butter				
Margarine				
Shortening				

CHARACTERISTICS OF A HIGH-QUALITY BAKED CAKE

Color:	*Golden brown.*
External appearance:	*High with slightly rounded, smooth top.*
Internal appearance:	*Fine, even texture, not crumbly.*
Body:	*Soft, velvety, slightly moist, light, tender.*

QUESTIONS

1. What role does flour play in the cake?

2. How would you substitute all-purpose flour for the cake flour in the recipe?

3. What role do sugar and fat play in the cake recipe?

4. Why would shortening cream be better than butter or margarine?

5. Why should the egg, butter, and margarine be at room temperature before incorporation into the cake batter?

6. What are the leavening agents in the creaming-style cake?

7. How can you test when the baked cake is ready to be removed from the oven?

8. How are layer cake pans positioned in the conventional oven? Do you follow the same protocol for a convection oven?

9. Which cake was preferred in flavor and why?

B. ONE-BOWL METHOD (QUICK MIX METHOD)

1. Yellow Cake

1 cup cake flour
2/3 cup granulated sugar
1/4 teaspoon salt
1½ teaspoons double-acting baking powder

1/4 cup shortening, butter, or margarine**
1/2 cup milk
1/2 teaspoon vanilla extract
1 large egg, room temperature

1. Preheat oven to 350°F. Set rack in the middle position of the oven.
2. Cut waxed paper to fit the bottom of an 8-inch round cake pan.
3. Grease only the bottom of the cake pan; insert the waxed paper; grease the paper again.
4. Sift together in a medium-sized mixing bowl, the flour, sugar, salt, and baking powder.
5. Add the shortening, milk, and vanilla. Using a hand mixer, start at low speed to moisten the flour for about 30 seconds; then at medium speed beat the mixture for **2 minutes, scraping the bowl constantly**.
6. Add the egg and beat the mixture for **2 minutes, scraping the bowl constantly**.
7. Push the batter at one time into the prepared cake pan. Level the batter evenly. Bake for approximately 25 minutes. Test for doneness.
8. Remove pan from oven, and allow the cake to cool in an upright position for 10 minutes. Then remove the cake from the pan and place on a wire rack to cool.

Each unit should be assigned a specific fat. Butter and margarine should be removed from the refrigerator and brought to room temperature prior to use. When the cakes are ready, each student should sample the finished product, and record the results in Table 5.3.

Table 5.3	TABLE FOR EVALUATION OF SHORTENED-STYLE CAKE BY THE ONE-BOWL METHOD			
Fat	Crust Color	Grain	Moistness	Flavor
Butter				
Margarine				
Shortening				

QUESTIONS

1. Why is it important that the butter or margarine be at room temperature for the one-bowl method when no creaming of the fat was involved?

2. Why does the one-bowl method cake recipe contain more baking powder than the creaming method recipe?

3. How do cakes made with the one-bowl method compare with cakes made with the creaming method?

Table 5.4 FLAWS IN BAKED SHORTENED-STYLE CAKES AND THEIR POSSIBLE CAUSES	
What Happened If?	Possible Causes
Cake does not rise properly	Too much liquid or shortening; too large a pan; too cool an oven
Cake is peaked or cracked	Too much flour; too hot an oven
Cake is pale	Too little sugar; too short baking time
Cake is coarse grained	Too much shortening; underbeaten
Cake is crumbly	Too much sugar or shortening; underbeaten
Cake is dry	Too much baking powder; too long baking time
Cake is heavy	Too much liquid, shortening or flour

II. TO DEMONSTRATE THE EFFECT OF CHANGING THE pH IN CHOCOLATE CAKE BATTERS

A. RED DEVIL'S FOOD CAKE

1 cup cake flour
1/4 cup shortening
3/4 cup granulated sugar
1/2 cup buttermilk
3/4 teaspoon baking soda

1/2 teaspoon vanilla extract
1/2 teaspoon salt
1 large egg, room temperature
1/4 cup cocoa powder

1. Preheat the oven to 375°F. Set rack on the middle position in the oven.
2. Lightly grease the bottom of a 9-inch round layer pan. Insert with a cut round of waxed paper; lightly grease the waxed paper; evenly shake over the greased bottom 1 tablespoon of flour. Toss out any excess flour left on the bottom of the pan. Sift together the flour, sugar, soda, salt, and cocoa into a medium-sized mixing bowl.
3. Add to the dry ingredients, shortening, buttermilk, and vanilla.
4. Beat the ingredients on low speed for 30 seconds until moistened; then beat at medium speed for 2 minutes, scraping the bowl constantly.
5. Add egg to the batter. Beat for 2 minutes, scraping the bowl constantly.
6. Remove 1 tablespoon of cake batter and place it in a custard cup; add 1 tablespoon distilled water; mix. Measure pH: ____.
7. Use a rubber scraper to push all the batter into the prepared pan. Level the batter in the pan.
8. Bake for 25–30 minutes or until cake tests done.
9. Cool in upright position for 10 minutes before removing. Remove cake from pan and cool on a wire rack.
10. Record observations in Table 5.5.

B. VARIATION 2 (pH < 7)

Follow Devil's Food Cake Recipe, except:

Use $1\frac{1}{4}$ teaspoons baking powder in place of the baking soda.
Record observations in Table 5.5.

C. VARIATION 3 (pH > 7)

Follow Devil's Food Cake Recipe, except:

Use 1/2 cup whole milk, instead of the buttermilk, and increase the baking soda to $1\frac{1}{4}$ teaspoons.
Record observations in Table 5.5.

Table 5.5 TABLE FOR THE EVALUATION OF CHOCOLATE CAKES				
Liquid and Leavening Used	pH of the Batter	Cell Size of the Cake	Flavor of the Cake	Color of the Cake
Devil's food cake				
Variation 2				
Variation 3				

QUESTIONS

1. How did the pH variations affect the color of the chocolate cake?

2. How was the crumb of the chocolate cake affected by

   a. increasing the pH?

    b.   decreasing the pH?

3.   How was the flavor of the chocolate cake affected by

    a.   increasing the pH?

    b.   decreasing the pH?

4.   What is European-style cocoa and how would it affect the quality of the chocolate cake if it was substituted for natural cocoa?

III. FOAM-STYLE CAKE

To learn and observe the stages that egg whites are to be whipped and the proper mixing and folding techniques to make a proper foam-style cake.

A. <u>ANGEL FOOD CAKE</u>

1 cup sifted cake flour
$1\frac{1}{2}$ cups granulated sugar
1/2 teaspoon salt
$1\frac{1}{2}$ cups egg whites, room temperature
  (about 12 large eggs)

$1\frac{1}{2}$ teaspoons cream of tartar
$1\frac{1}{2}$ teaspoons vanilla extract
1/2 teaspoon almond extract

1.   Preheat oven to 375°F. Set oven rack to the lowest position in the oven.
2.   Sift flour, salt, and 3/4 cup sugar together onto a large piece of waxed paper; repeat three times; set aside.
3.   Beat the egg whites to the foam stage in a large mixing bowl; add the cream of tartar, vanilla extract, and almond extract and beat the egg whites to the soft peak stage.
4.   Gradually add the remaining 3/4 cup sugar, 1 tablespoon at a time, until the sugar is added. Egg whites should be at the stiff peak stage. Rub some of the beaten egg whites between your fingers to determine if the sugar has dissolved. If not, continue beating, but check again and often.
5.   Sift approximately 1/3 of the flour mixture over the egg white meringue, and carefully fold in the flour, being careful to retain as much air as possible. Continue this process two more times working quickly, but carefully using your rubber spatula (**do not use an electric mixer**).
6.   Place batter in an **ungreased** 9-inch tube pan. Cut through the batter with a flat knife or long spatula to break up any large air pockets or large bubbles.
7.   Bake in a preheated oven for 40–45 minutes. Remove from oven when cake tests done, springs back when touched with a finger. Remove cake from oven and cool upside down in the pan until the cake is cooled for 2–3 hours.
8.   Record observations in Table 5.6.

B. CHOCOLATE CHIP ANGEL FOOD CAKE

1 cup sifted cake flour
1½ cups confectioner's sugar
3/4 cup granulated sugar
1/4 cup mini semisweet chocolate chips
1½ cups egg whites, room temperature
  (about 12 large eggs)

1¼ teaspoons cream of tartar
1/4 teaspoon salt
1½ teaspoons vanilla extract
1/2 teaspoon almond extract

1. Preheat oven to 375°F. Set oven rack to the lowest position in the oven.
2. On a sheet of waxed paper sift together cake flour and confectioner's sugar three times; set aside. In a small bowl, combine 2 tablespoons of the flour–sugar mixture with the chocolate chips; set aside.
3. In a large bowl of an electric mixer, beat egg whites to the foamy stage; then add the cream of tartar, salt, vanilla extract, and almond extract.
4. Continue beating the egg whites, and beat to the soft peak stage; gradually add granulated sugar, 1 tablespoon at a time, until stiff but shiny peaks are formed. Check the meringue by rubbing a small amount between your fingers and if it feels gritty, continue beating. Check again after another minute.
5. Gradually sift the flour–confectioner's sugar mixture (in three additions) over the meringue and carefully fold in. Sprinkle the chocolate chips over the meringue and fold in.
6. Carefully guide the batter into an ungreased 10-inch tube pan and bake for 30–35 minutes or until the cake tests done by pressing with your finger and the cake springs back.
7. Cool the cake upside down in the pan for at least 2 hours. Loosen the sides with a long slender spatula and remove.
8. Top with Chocolate Glaze (recipe below).

CHOCOLATE GLAZE

1/2 cup semisweet chocolate chips
1/4 cup butter or margarine, softened

2 tablespoons white corn syrup

In a small microwavable safe bowl add chocolate chips, butter, and corn syrup. At HIGH power heat ingredients for 30 seconds. Remove from microwave, stir; return, if necessary, to microwave and heat for 15 seconds; stir. Continue heating and stirring if needed until glaze is smooth. With a thin spatula spread the glaze over the top of the cake, allowing it to drip down the sides.

C. SPONGE CAKE FOR BEGINNERS

6 large egg yolks
1½ cups granulated sugar
1½ cups cake flour
1 teaspoon double-acting baking powder
1/2 teaspoon salt
1/3 cup cold water

1 teaspoon grated lemon or orange rind
1 teaspoon lemon or orange extract
1 teaspoon vanilla extract
6 egg whites (3/4 cup), room temperature
1/2 teaspoon cream of tartar

1. Preheat oven to 325°F. Set the oven rack to the lowest position.
2. Beat in a medium-sized bowl until thick the egg yolks and sugar, about 5 minutes.
3. Sift together the flour with the salt and baking powder. Beat into the yolk–sugar mixture the cake flour mixture alternately with the water, extracts, and grated rind.
4. In a large mixing bowl beat together the egg whites with the cream of tartar, until the peaks are stiff and shiny.
5. Take a small amount (1/4 cup) of the beaten egg white foam and stir it into the egg yolk mixture. Then gradually and gently cut and fold the egg yolk mixture into the egg white foam.
6. Pour batter into an ungreased 10-inch tube pan. Bake for 60–65 minutes, or until cake tests done when cake springs back when lightly pressed with your finger.
7. Remove cake from oven and turn upside down to cool for at least for 2 hours.
8. Record observations in Table 5.6.

D. JELLY ROLL CAKE (WHOLE EGG SPONGE CAKE)

1 cup cake flour
1 teaspoon double-acting baking powder
1/4 teaspoon salt
3 large eggs, room temperature
1 cup granulated sugar

1/3 cup water
1 teaspoon vanilla extract
1 teaspoon grated lemon or orange rind
1/2 cup sifted confectioner's sugar
3/4 cup strawberry preserves

1. Preheat oven to 375°F. Set the rack on the middle position.
2. Grease the bottom and sides of a jelly roll pan, 15$\frac{1}{2}$ × 10$\frac{1}{2}$ inches. **Line the bottom of the pan only with waxed paper or parchment paper. Grease the paper**.
3. Sift together the flour, salt, and baking powder; set aside.
4. Beat eggs in a small mixing bowl (1 quart) at high speed for 5 minutes. **This is very crucial to the recipe**. Mixture must be thick and triples in bulk.
5. Pour beaten egg foam into a larger bowl. Gradually beat in sugar.
6. Beat in water, vanilla, and lemon rind.
7. Gradually add flour mixture in three additions on low speed. Do not over beat.
8. Pour mixture into prepared pan. Level with a rubber spatula. Tap the pan gently on the counter to release any air bubbles.
9. Bake cake for 12–15 minutes. Cake is ready when it springs back when lightly touched.
10. While cake is baking, sift confectioner's sugar onto a linen towel about the size of the pan.
11. Loosen edges of cake; turn out onto confectioner's sugar. Remove paper. Trim off any crusty edges or sides.
12. While cake is hot, start at narrow end and roll cake and towel together. Allow cake to cool in this position on a cooling rack.
13. When the cake is cold, carefully open towel and unroll. Carefully spread jelly on the cake, leaving 1 inch space on each end.
14. Carefully reroll the cake. Dust lightly with confectioner's sugar. Cut into 1 inch slices for serving.
15. Record observations in Table 5.6.

Variations

a. 3/4 cup raspberry preserves could be substituted for the strawberry preserves.
b. Spread 1/2 cup lemon curd on the cake. Over the lemon curd spread 3/4 cup of raspberry preserves.

Table 5.6 TABLE FOR THE EVALUATION OF FOAM-STYLE CAKES					
Cake	Appearance of Batter	Size of the Cake	Grain of the Cake	Texture of the Cake	Flavor of the Cake
Angel food cake					
Sponge cake					
Jelly roll cake					

QUESTIONS

1. In making a sponge or angel food cake, what manipulative technique affects:

   a. large volume?

   b. a fine grain?

   c. a coarse texture?

2. What role does cream of tartar play in the foam-style cake?

3. Why was part of the sugar added to the egg white foam in the angel food cake?

4. What is the tenderizing agent(s) in the angel food and sponge cakes?

5. Why should the foam-style cakes be baked as soon as they are mixed?

6. What precautions should be followed when adding flour to egg whites foams?

7. Why was the tube pan ungreased for the sponge cake and angel food? What would have happened if the sides were greased?

8. Discuss and compare the nutritional quality of angel food cake, sponge cake, and shortened-style cake.

LABORATORY 6

Pastry, Cream Puffs, and Popovers

LABORATORY 6
Pastry, Cream Puffs, and Popovers

Pie dough is a simple product made up of flour, water, salt, and fat. The selection and manipulation of ingredients will have a strong effect on the tenderness and flakiness of the pie pastry. The leavening agent is steam which will contribute to flakiness. Steam is also responsible for the leavening effect observed in cream puffs and popovers. Therefore, the student will study pies, cream puffs, and popovers in this laboratory exercise and how manipulation of the ingredients and steam will affect the final quality of these baked products.

VOCABULARY

cut in
flakiness
lard

pastry blender
pastry method

plastic fat
tenderness

OBJECTIVES

1. To observe how steam plays a common role in three different products: pie dough, popovers, and cream puffs.
2. To learn that proper manipulation will have a strong effect on the outcome of the pie dough, cream puffs, and popovers.
3. To determine which fat contributes to both tenderness and flakiness in a pastry dough.
4. To distinguish between flakiness and tenderness in pastry dough.

PRINCIPLES

1. Steam is the leavening agent in pie crust, popovers, and cream puffs; therefore, a hot oven is needed.
2. Cream puffs are made from a dough and popovers are made from a batter, but both products have the same ratio of liquid to flour (1:1).
3. Popover batter is made by the muffin method: liquid ingredients added to the dry ingredients.
4. Cream puffs are made by a unique mixing method that involves the gelatinization of starch and addition of the egg to provide structure and emulsification.
5. The pastry method involves cutting the fat into the flour.
6. The type of fat selected for the pastry dough will determine the flakiness and tenderness of the pie crust.
7. Flakiness and tenderness are not equal to each other, and are affected by different entities.
8. Too much water and over manipulation cause gluten to develop and will toughen the pie crust.

I. TO LEARN AND OBSERVE HOW MANIPULATION AND STEAM WILL AFFECT CREAM PUFFS AND POPOVERS

A. CREAM PUFF SHELLS

1/4 cup water
1/4 cup all-purpose flour
2 tablespoons butter

1 1/2 teaspoons sugar
Dash salt
1 large egg, beaten

1. Preheat oven to 400°F. Lightly grease three areas of a baking sheet, each area approximately 2 inches in diameter. Allow approximately 3 inches between greased areas.
2. Place water, butter, and salt in a small-sized saucepan; heat until the butter is melted and water boils vigorously.
3. Add the flour and sugar all in one portion to the boiling water–fat mixture. Stir quickly with a wooden spoon to get flour well blended with the water–fat mixture. Stir mixture until it forms a ball. Cook for 1 minute over medium heat; remove saucepan from heat.
4. Cool the cooked starch paste slightly.
5. Add the egg to the starch paste and beat vigorously (use a wooden spoon or a hand mixer). **This step is important. The starch paste must be sticky at this point. If the paste has a greasy feel it is not**

correct. **Beat another egg and add it gradually to the paste. The mixture should be sticky at this point. Therefore,** *beating* **and the** *egg* **provide the emulsion needed to form the proper dough.**

6. Divide paste mixture into three equal portions. Place each portion on the greased area on the baking sheet.
7. Bake for 35 minutes. Check after 25 minutes.
8. Remove baked shells from baking pan and cool on wire rack. Fill with creamy vanilla pudding.
9. Record observations in Table 6.1.

Table 6.1 TABLE FOR THE EVALUATION OF CREAM PUFFS AND POPOVERS				
Product	Observation of Batter or Dough	Crust Color	Flavor	Volume
Cream puffs				
Popovers				

B. CREAMY VANILLA PUDDING

1½ cups skim milk
1 egg yolk
1/3 cup granulated sugar
1 tablespoon butter

1/8 teaspoon salt
1½ teaspoons vanilla
2 tablespoons + 2 teaspoons cornstarch

1. Blend cornstarch, sugar, and salt in a 1-quart saucepan.
2. Mix together milk and egg yolk with a whisk. Blend well and make sure there is no trace of yolk pieces floating in the milk.
3. Add the milk–yolk mixture to the dry ingredients and mix until the dry ingredients are dissolved.
4. Place saucepan over medium heat and stir mixture constantly with a wooden spoon. Bring mixture to a boil, and boil for 1 minute, stirring constantly. Remove from heat.
5. Add butter and vanilla; stir to incorporate ingredients.
6. Cool filling thoroughly before filling cream puff shells. To prevent skin from forming on the surface of the pudding, place a piece of plastic wrap directly touching the surface of the warm pudding. **Variation: whip 1/4 cup of heavy cream to form stiff peaks. Fold into the** *cold pudding.*

C. POPOVERS

1/2 cup all-purpose flour
1/4 teaspoon salt
1/2 cup milk

1 large egg
Vegetable spray for custard cups or popover tin

1. Preheat the oven to 450°F. Set oven rack to the lowest position.
2. Thoroughly grease bottom and sides of 3 or 4 custard cups, popover pans, or deep aluminum muffin tins.
3. Sift together flour and salt into a 1-quart mixing bowl.
4. Add egg to milk and mix thoroughly.
5. Add the milk/egg mixture to the flour mixture and use a rotary mixer to blend the wet and dry ingredients. Beat until mixture is smooth.
6. Fill custard cups or popover pans 1/3–1/2 full. Muffin tins are to be 1/2 full.**
7. If using custard cups, place on a baking sheet. Place in the oven.
8. Bake for 20 minutes; **reduce oven temperature to 350°F** for 20 minutes. Remove popovers from the oven; place a small slit in the top of the popover and place the popovers back into the oven **which has been turned off** and allow them to stay in the oven for 5 minutes.
9. Record observations in Table 6.1.

**HINT: If using popover pans, preheat the pans while preparing the batter. This will help to create the steam and bigger popovers.

1. What is the leavening agent for the popovers and cream puffs?

2. Why is there a difference in the browning reaction between the cream puffs and the popovers?

3. Why would over manipulation not be as critical in the cream puff dough as opposed to the popover batter?

4. What function does the egg perform in the cream puff dough?

5. If the cream puff dough appears greasy after the egg has been added, how may the situation be corrected to save the recipe?

6. Why is there a difference in texture between the cream puffs and the popovers?

II. TO DETERMINE WHICH FAT CONTRIBUTES BOTH TENDERNESS AND FLAKINESS IN A PIE CRUST THROUGH PROPER MANIPULATION

A. PIE CRUST (BASIC RECIPE)

1 cup all-purpose flour	1/3 cup butter, margarine, or shortening
1 teaspoon salt	2–3 tablespoons cold water

1. Stir together in a medium-sized bowl the flour and the salt.
2. Add the fat to the dry ingredients and with a pastry blender cut the fat into the flour until the mixture looks like it contains large peas.
3. Sprinkle the water over the mixture and toss the mixture with a fork until it starts to come together. **If the mixture fails to come together, add more water, a small amount at a time.**
4. With the tips of your fingers gather the dough into a ball.
5. Wrap the dough in plastic wrap and place it in the freezer for 10 minutes to allow the dough to relax. This allows for more hydration and also relaxes the dough for easier rolling.

Rolling of the Dough (Fig. 6.1)

1. Place the dough between two large pieces of waxed paper for easier rolling and handling.
2. Place the bottom sheet on the work counter; sprinkle generously with flour; take your dough and gently roll it around in the flour; take your second sheet of waxed paper and place it on the dough, pressing down on the dough ball with the palm of the hand until it is a slightly flattened disc.
3. Place the rolling pin on the center of the mound of dough and start rolling from the center outward. **Never run the rolling pin back and forth over the dough as it will only toughen the pastry.**

FIG. 6.1: Roll the pastry dough to 1/8 inch thickness from the center outward to the edge.

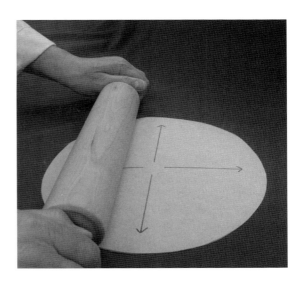

4. The dough should be rolled out to 1/8 inch thickness and it should be slightly larger than the pie pan. Remove the top piece of waxed paper.
5. Loosen pastry dough from the waxed paper (run a thin spatula underneath the dough) with minimum stretching of the dough.
6. Ease the dough into the pie pan, but be careful not to stretch the dough. Gently press the pastry dough against the bottom and the sides of the pie pan. Make sure that the pastry overhangs the sides of the pie pan. This is important to crimp the edges or if for a double crust pie to seal the top crust.
7. If the shell is going to be baked without a filling, "baking the crust blind," prick the bottom and sides of the crust to allow the steam to escape. **Another variation: line the bottom of the crust with parchment paper and weigh dough the crust with dried beans if the crust is going to be partially baked.**
8. Bake the pricked pie crust at 475°F for 8–10 minutes until a *light* brown. The darker the crust more fat is broken down.
9. **Summary:**
    a. Crust will have a blistered appearance which is a sign of flakiness.
    b. A light colored crust is desirable; too dark will cause excessive fat breakdown.
    c. If the crust is not pricked sufficiently, puffing will occur because the steam was not allowed to escape.
    d. Just because the crust is flaky is not an indication that it is tender. Tenderness is a result of
        i.   low amount of handling,
        ii. the correct type of fat,
        iii. the correct amount of water.

B. LOW-FAT PIE CRUST

1¹/₄ cups all-purpose flour   1/4 cup shortening
1/4 teaspoon salt   4–5 tablespoons cold water

1. Sift together flour and salt into a medium-sized bowl.
2. Cut in shortening until the mixture resembles cornmeal.
3. Sprinkle water over the mixture and toss the water and flour with a fork until the mixture starts to come together and adheres in large clumps.
4. Using the tips of your fingers, gather the dough into a ball.
5. Cover the dough in plastic wrap and refrigerate for 10 minutes.
6. Follow the instructions for the rolling and baking of the dough.

C. PIE DOUGH WITH OIL

1¹/₄ cups all-purpose flour   1/4 cup vegetable oil (such as soybean oil, corn oil)
1/4 teaspoon salt   4–5 tablespoons milk

1. Sift together flour and salt into a medium-sized bowl.
2. Add oil and toss with a fork to combine.
3. Add enough milk to hold dough together. **This is a tricky step and it may seem that you have added enough liquid. If not enough milk is added, when the dough is rolled out it will be crumbly and fall apart. Therefore, it is better to add more than less milk.**
4. Follow the steps concerning the rolling and baking of the dough.

D. PIE WAFERS

1. Prepare Pie Crust (Basic Recipe) above, but each unit should use a different fat. Allow the dough to rest for 10 minutes before rolling.
2. Cover the surface with waxed paper and place the pastry guides (1/4 inch thick) vertically and parallel to each other 4 inches apart on the covered surface.
3. Place the dough between the pastry guides. Use your hands to pat out the dough into an oblong shape approximately 1 inch thick.
4. Cover the dough with a piece of waxed paper.
5. Lightly roll the oblong-shaped dough with the rolling pin to obtain a uniform 1/4 inch thickness throughout the dough. **Be sure that the product is not on the wood guides after shaping is completed. This is important so that the dough thickness is controlled.**
6. Gently peel off the upper layer of waxed paper. Cut the pastry into uniform-sized wafers (1½ × 2½ inches).
7. Bake at 450°F for approximately 8–10 minutes. Watch the time carefully and remove the wafers when a very light brown.
8. Cool; place on racks to cool thoroughly. Evaluate the wafers for flakiness and tenderness.
9. Record observations in Table 6.2.

Table 6.2 TABLE FOR EVALUATION OF PASTRY WAFERS				
Shortening Used	Blistering Effect	Flakiness*	Tenderness**	Flavor
Shortening				
Lard				
Butter				
Margarine				
Oil				

*Flakiness can be described as: 1. very thick layers; 2. moderately thick layers; 3. slightly thick/thin layers; 4. moderately thin layers; 5. very thin layers.

**Tenderness can be described as: 1. very tough/crumbly; 2. moderately tough/crumbly; 3. slightly tough/crumbly; 4. tender; 5. very tender.

CHARACTERISTICS OF HIGH-QUALITY PIE PASTRY

Color: Light brown.
Appearance: Blistered on top.
Tenderness: Cuts easily and holds its shape.
Flakiness: Has layers that can separate easily.

---

1. a. Which fat produced the most tender crust? Why?

    b. The most flaky crust? Why?

    c. The most flaky and most tender crust? Why?

2. What affect does manipulation have on the pie crust?

3. a. What affect would too much water have on the pie crust?

    b. Too little water?

4. Why was the pastry dough pricked with a fork before baking?

5. Why is all-purpose flour used in the pastry dough? Could you substitute cake flour in the pastry dough recipe?

6. Why is oil used at a lower rate in the pastry dough recipe? Could the oil recipe be considered a low-fat pastry dough and why?

E. APPLE PIE

6 medium apples (Gala, or Golden Delicious, or Rome
   Beauty, or an equal combination of all three apples)
1/2 cup sugar
Dash salt
2 tablespoons flour

1/4 teaspoon nutmeg
3/4 teaspoon cinnamon
2 tablespoons butter or margarine
Pastry for 9 inch, 2 crust pie (double crust recipe)

1. Preheat oven to 425°F.
2. Peel apples and slice thin. Place apples in a large bowl.
3. Mix together the sugar, salt, flour, nutmeg, and cinnamon. Sprinkle mixture over the sliced apples and toss to coat.
4. Place apples into pastry lined pan. Place pieces "dot" of butter over the fruit. Cover fruit with pastry; crimp the pastry to seal in contents. Make slits on top of crust to allow steam to escape.
5. Bake for 30–35 minutes or until filling starts to bubble through the slits. Remove pie from oven and allow to rest before serving. Can be served warm or cold.

6. **To avoid excessive browning of the crust edges, put narrow pieces of aluminum foil around the edges. Leave them on for the first 20 minutes of the baking, then remove and continue baking the pie.**

F. PEAR PIE

Follow directions for Apple Pie, except use 6 cups thinly sliced pears, such as bartlett or anjou.

G. PEACH PIE

2 (1 pound) cans sliced peaches, well drained
2 teaspoons lemon juice
1/4 teaspoon almond extract
1/2 cup light brown sugar, packed
2 tablespoons flour

1/2 teaspoon cinnamon
1/4 cup syrup from canned peaches
2 tablespoons butter or margarine
1 teaspoon granulated sugar
Pastry crust for 9-inch 2 crust pie

1. Heat oven to 425°F. Prepare pastry.
2. Drain peaches and save juices.
3. Toss drained peaches with lemon juice and almond extract.
4. Stir together brown sugar, flour, and cinnamon. Toss sugar mixture with the peaches.
5. Turn mixture into pastry lined pan. Pour 1/4 cup of the drained juice over the peach mixture.
6. Dot with butter. Cover with top crust; seal and flute. Make slits on top. Sprinkle with sugar.
7. Bake 35–45 minutes or until juices bubble through the slits.

H. CREAM PIE FILLING

1/2 cup granulated sugar
1/2 teaspoon salt
4 tablespoons cornstarch
2 cups 2% milk

2 large egg yolks
2 teaspoons butter or margarine
1 teaspoon vanilla extract
1 baked 9-inch pie crust

1. Mix together sugar, salt, and cornstarch in a medium-sized saucepan.
2. Blend thoroughly with a whisk the milk and egg yolks; stir this into the dry ingredients. Mix well until all ingredients are dissolved.
3. Cook mixture over medium heat; stirring constantly. Bring mixture up to a boil and boil for 1 minute, stirring constantly.
4. Remove thickened mixture from the heat and stir in vanilla extract and butter. Stir until butter is melted. Pour mixture into baked pie shell, and allow mixture to cool. Then refrigerate pie until serving time.

Variations

1. *Banana Cream Pie*: Add one sliced ripe banana to the baked pie shell before pouring in the cooled filling. Sweetened whipped cream can serve as a topping or following directions for a **meringue topping** (Meringue).
2. *Coconut Cream Pie*: Add 1/3 cup grated sweetened coconut plus 1/4 teaspoon almond extract in Step 4 of the Cream Pie Filling. Can be served with sweetened whipped cream as a topper or pile on meringue and baked as directed (**Meringue**). Sprinkle sweetened coconut on the meringue before baking.
3. *Chocolate Cream Pie*: Increase the sugar to 3/4 cup; add $1\frac{1}{2}$ squares unsweetened chocolate (finely chopped) to the milk and dry ingredients. Proceed as in the Cream Pie Filling. Serve topped with sweetened whipped cream with shaved chocolate or a **Meringue**.

I. LEMON MERINGUE PIE

1/2 cup granulated sugar
$2\frac{1}{2}$ tablespoons cornstarch
1/8 teaspoon salt
1 cup 2% milk
2 large eggs, separated

1/2 teaspoon finely grated lemon peel
$2\frac{1}{2}$ tablespoons fresh lemon juice
1 tablespoon butter or margarine
1 baked pie shell (9-inch pie shell)

1. Combine sugar, cornstarch, and salt in a medium saucepan.
2. Whisk together the milk and egg yolks. Add to the dry ingredients and blend well. Take care to dissolve all lumps.
3. Cook over medium heat stirring constantly. Bring mixture to a boil and boil for 1 minute stirring constantly.
4. Remove mixture from the heat and slowly and gradually stir in the lemon juice. Add the butter and the lemon rind; mix well.
5. Pour filling into baked pie shell; top with Meringue (recipe below) and bake.

Meringue

3 large egg whites
1/4 teaspoon cream of tartar

1/2 teaspoon vanilla extract
$4\frac{1}{2}$ tablespoons granulated sugar

1. Preheat the oven to 350°F. Set the rack to the middle position.
2. Beat the egg whites in a medium-sized bowl to the foam stage. Add cream of tartar and vanilla.
3. Beat until the egg whites form soft peaks. Gradually add the sugar, 1 tablespoon at a time. Beat until the meringue is shiny, forms stiff peaks, and when a small amount of the meringue is rubbed between the fingers it should feel smooth and not gritty.
4. Place the meringue on top of the hot filling, making sure that the meringue touches the crust; this acts as a sealant and prevents shrinkage of the meringue during baking.
5. Bake the pie with meringue in the preheated oven for 12–15 minutes. (**The rule of thumb when baking a meringue is that for every egg white used the time is 5 minutes per egg white.**)

J. QUICHE LORRAINE

Pastry for 9-inch pie (the oil crust works nice for this recipe)
2 tablespoons vegetable oil, divided
6 slices of turkey bacon
1 cup (4 oz.) natural Swiss cheese or Guyere cheese, shredded
1/3 cup finely minced onion

2 eggs
2 egg whites or 1/2 cup egg substitute
2 cups 2% milk
2 tablespoons cornstarch
1/2 teaspoon salt
1/4 teaspoon black pepper
1/8 teaspoon ground red pepper

1. Heat oven to 425°F. Prepare pastry, roll out, and set aside.
2. In a frypan, add 1 tablespoon of the vegetable oil; heat oil; add turkey bacon and cook until the bacon is crisp; remove from pan and place bacon strips on paper towel. Allow the bacon to cool. Once cool, crumble the bacon.
3. Add the remaining tablespoon of the vegetable oil to the frypan; over moderate heat add the chopped onion, and cook the onion until tender and translucent, but not brown. Remove the onions from the heat.
4. In a 1-quart measuring cup, mix together the eggs, egg whites, milk, cornstarch, salt, black pepper, and ground red pepper. Make sure all ingredients are blended well.
5. In the pastry lined pan add the crumbled bacon, cooked onion, and the shredded Swiss cheese. Carry the pie pan to the oven and place the pie dish on the oven rack.
6. Carefully pour the egg/milk mixture into the pastry lined pan. Bake the pie for **15 minutes. Reduce the oven temperature to 350°F and bake until a knife comes out clean (about 25 minutes).** The center of the filling may seem a little loose, but it will set as it cools.
7. Cool the pie for 10 minutes before serving.

K. CHICKEN POT PIE

1 package (10 oz.) frozen peas and carrots
3 tablespoons butter or margarine
1/3 cup all-purpose flour
1/3 cup chopped onion
1/2 teaspoon salt

1/4 teaspoon black pepper
$1\frac{3}{4}$ cups chicken broth (see below)
2/3 cup skim milk
$2\frac{1}{2}$–3 cups diced cooked chicken
Pastry for 9-inch double crust pie

Broth

3 cups water
2 medium-sized boneless, skinless
   chicken breasts
2 onions, peeled and quartered
2 celery stalks, quartered

1/4 teaspoon ground thyme
3 chicken bouillon cubes
Dash salt and pepper
Place ingredients in a large saucepan. Bring to a
   boil; cover and simmer for 1 hour.

1. Preheat oven to 425°F.
2. Rinse frozen peas and carrots under cold running water to separate; drain.
3. Melt butter in a 2-quart saucepan over low heat until melted. Add onion and cook until the onion is soft and translucent; add pepper and salt. Add flour and mix (a paste will form).
4. Remove pan from heat; whisk in broth and milk until the paste is dissolved. Return the pan back to the heat. Stir constantly until the sauce thickens and comes to a boil; boil and stir for 1 minute. Stir in chicken and vegetables.
5. Pour filling into pastry lined pan; place the other part of the crust on top of the filling. Seal and crimp the crust. Place slits on the top crust.
6. Bake at 425°F for about 30–35 minutes or until filling starts to bubble through the slits in the crust. Cool for 10 minutes before serving.

QUESTIONS

1. Why are slits made in the top crust of a 2-crust pie?

2. How can you prevent the edges of the pie crust from burning or becoming excessively brown?

3. Why is the lemon juice added at the end of the cooking period when making the lemon meringue filling?

4. Describe the procedure on adding the meringue on a pie.

5. What material should a pie pan be made of to insure an evenly baked pie with a crisp undercrust?

6. Why are the temperatures high when baking a pie crust?

7. What are the nutritional consequences in selecting pies as part of the diet?

8. How can a pie be used as part of a meal?

9. The directions state that the meringue pie is baked on the middle rack. Some cookbooks give the direction to use the top rack as a choice. What precaution should you follow if using the top rack in the oven?

LABORATORY 7

Fruit Selection and Cookery

LABORATORY 7
Fruit Selection and Cookery

There are many varieties of fruit that can be found in the market. This laboratory exercise is designed to demonstrate to the student how selection of a particular variety of fruit will affect its preparation.

VOCABULARY

antioxidant
cellulose
dried fruit
hemicellulose
modified atmospheric storage

osmosis
pectic acid
pectin
protopectin

polyphenoloxidase
rehydrate
semipermeable membrane
sulfur dioxide

OBJECTIVES

1. To study the prevention of browning in fresh fruit through the use of antioxidants.
2. To emphasize that different varieties of the same fruit will not have the same cooking properties.
3. To observe the effect of moist or dry heat and sugar on cooking fruit.
4. To study the principle of osmosis and its effect on the structural properties of fruit during cooking.
5. To learn the proper way to rehydrate and cook dried fruit.

PRINCIPLES

1. Osmotic pressure causes:
    a. rupture of the cell's structure when cooked in water,
    b. retention of cell structure when cooked in sugar syrup.
2. Fruit can be cooked in a variety of ways: poached, baked, broiled, sautéed; but not all varieties of the same kind of fruit can be cooked in the same way.
3. The two main pigments found in fruit are carotenoids and anthocyanins.
4. When certain fruits are cut, darkening takes place caused by enzymatic browning. Antioxidants are used to prevent this darkening.
5. During the ripening process, protopectin is changed to pectin. Pectin in the presence of certain amount of sugar and acid will form a gel (jelly).
6. Overripe fruit contains pectic acid from the breakdown of pectin. Fruit will have a mealy and mushy texture.
7. Ripeness stage of the fruit will affect its cooking quality.

I. TO STUDY THE PREVENTION OF BROWNING IN FRESH FRUIT THROUGH THE USE OF ANTIOXIDANTS

A. BROWNING OF FRUIT

1. Slice a banana, pear, apple, or avocado and divide it into six lots.
2. Treat the six lots as indicated below.
    Treatment No. 1: Air.
    Treatment No. 2 : Cream of tartar solution (1/8 teaspoon + 1/2 cup water).
    Treatment No. 3: Lemon juice solution (1/8 teaspoon + 1/2 cup water).
    Treatment No. 4: Ascorbic acid solution (1/8 teaspoon + 1/2 cup water).
    Treatment No. 5: Salt solution (1/8 teaspoon + 1/2 cup water).
    Treatment No. 6: Blanch in boiling water for 3 minutes.
3. Record observations in Table 7.1 of any color change in the fruit at the end of 30 minutes.

Table 7.1 TABLE FOR THE EVALUATION OF FRUIT BROWNING			
Fruit	Treatment Number	Observation	Explanation

QUESTIONS

1. Describe the process of enzymatic browning.

2. List ways that enzymatic browning can be prevented and why is each one effective.

3. Explain why fruits brown when bruised, even when the skin is intact.

4. Why is that bananas mixed with cubed oranges in a fruit cup do not turn brown?

5. A catering company must set up a cheese and fruit tray for a party. The fruit that will be sliced are Red Delicious apples. Which treatment would work the best for keeping the fruit from turning brown during the party and why?

II. TO IDENTIFY THE FACTORS THAT INFLUENCE THE QUALITY OF COOKED APPLES

Select five cultivars of apples for each preparation (such as Red Delicious, Golden Delicious, Granny Smith, Rome Beauty, and McIntosh). Record all observations for each preparation in Table 7.2.

A. ORANGE HONEY BAKED APPLE

1 large apple	dash cinnamon
dash nutmeg	1 tablespoon honey
1/4 cup orange juice	

1. Preheat the oven to 375°F.
2. Rinse and core apple. Using a vegetable peeler, remove about a 1-inch strip of apple skin around the center of the apple.
3. Place apple in a 1-quart casserole dish.
4. Combine orange juice, honey, nutmeg, and cinnamon. Stir to dissolve the honey. Pour over the apple and cover the dish.
5. Bake apple for 25 minutes. Uncover and baste frequently with the juice in the pan for an additional 20–25 minutes. Remove from oven. Cool slightly.

B. APPLE SAUCE

2 medium apples	1/2 cup water
dash cinnamon	2 tablespoons granulated sugar

1. Wash, peel, and quarter apples. Core each quarter and cut into 3 or 4 slices.
2. To a small saucepan add water, cinnamon, and apples. Bring water to a boil, lower heat to the lowest setting and cover the pan. Simmer until the slices are tender, about 15–20 minutes.
3. Stir the cooked apples with a fork until the slices are broken up. Add sugar and stir until the sugar is dissolved.

C. CODDLED APPLE

1 medium-sized apple	1/2 cup granulated sugar
3/4 cup water	

1. Add water and sugar to a wide bottom frypan that has a cover. Stir the mixture until the sugar is dissolved.
2. Wash, peel, and core the apple. Place the apple on its side and slice the apple into 1/4 inch thick rings.
3. Add the apples to water–sugar mixture and bring it to a boil. Once the mixture boils, **cover it** and lower the heat so that syrup simmers and the apples cook slowly.
4. Cook the apples for about 20 minutes. Keep checking that syrup does not evaporate or if it gets too thick. If this happens, add a small amount of water to the pan. When the apples look clear and translucent remove from heat.

Table 7.2 TABLE FOR EVALUATION OF COOKED APPLES				
Variety	Cooking Style	Flavor	Texture	Appearance

QUESTIONS

1. Why does fruit soften when cooked in water?

2. Why are fruits cooked in syrup more translucent than when raw?

3. Explain the effect of the different concentrations of sugar on texture, shape, and fruit flavor.

4. What affect would the degree of ripeness have on the cooking quality of the fruit?

III. **TO OBSERVE FRUITS THAT HAVE ENOUGH PECTIN TO FORM A GEL WHEN SUGAR AND WATER ARE ADDED**

A. **CRANBERRY JELLY**

1 cup fresh cranberries   1/2 cup water
1/2 cup granulated sugar

1. Sort and wash cranberries. Place in a small saucepan.
2. Add water. Bring to a boil; boil for 5 minutes or until the skin has popped open.
3. Rub the cooked berries through a strainer or put through a food mill, collecting the juice and as much of the pulp as possible in a clean saucepan placed underneath.
4. Add sugar to the sieved pulp. Stir until the sugar is dissolved.
5. Heat the mixture to a boil, and boil for 5 minutes. Stir constantly to prevent mixture from sticking.
6. At the end of the cooking period, the mixture should be given a jelly "sheeting test": Several drops of the juice tend to flow together from the side of the spoon (cook until this point is reached).
7. Pour the jelly into a serving dish. Do not stir the jelly while it cools.

QUESTIONS

1. During the ripening of fruit, follow the development of pectin; what enzymes are involved?

2. What are the ingredients that are needed for the setting of the gel?

3. What acid is naturally found in cranberries?

IV. TO OBSERVE THE VARIETY OF WAYS FRUITS CAN BE PREPARED

A. TROPICAL AMBROSIA

2 mangos, peeled and cut into bite-sized chunks
2 large naval oranges, peeled
2 bananas, peeled and sliced thin
1 cup fresh pineapple, cut in thin wedges

1/2 honeydew melon, flesh cut into bite-sized chunks
1/3 cup confectioners' sugar, sifted
1 can (3$\frac{1}{2}$ oz.) sweetened flaked coconut
1/4 cup orange juice

1. Remove all outer white membrane from oranges and slice thin crosswise.
2. Layer oranges, bananas, mangos, honeydew, and pineapple in a serving bowl, sprinkling with sugar, coconut, and orange juice as you go.
3. Cover and chill for 2–3 hours. Mix lightly and serve.

B. BERRY CAKE

1$\frac{1}{2}$ cups assorted whole berries
    (such as, blackberries, blueberries, and raspberries)
1 cup all-purpose flour
3/4 cup + 1 tablespoon granulated sugar
1/2 teaspoon baking soda

1/2 teaspoon ground ginger
1/4 teaspoon salt
1/2 cup low-fat buttermilk
2 tablespoons vegetable oil
2 large eggs, beaten

1. Preheat oven to 375°F. Spray an 8-inch round pan with nonstick cooking spray. Line the bottom of the pan with waxed paper; spray again.
2. Place berries in a bowl with water and allow the berries to stand in the water for 1 minute. Drain the berries and place them onto a large piece of paper towel to dry.
3. In a medium-sized bowl, sift together the flour, 3/4 cup granulated sugar, baking soda, ginger, and salt. Make a well in the center of the flour and pour in the buttermilk, oil, and eggs. **Stir just until no dry flour is visible.**

4. Scrape the batter into the prepared pan. Spoon the berries on top and sprinkle with the remaining 1 tablespoon of sugar. Bake for 35–40 minutes or until a toothpick inserted in the center comes out clean. Cool on a wire rack. Remove from pan, transfer to a plate and serve.

C. MARBLED CHOCOLATE BANANA BREAD

$2^1/_4$ cups cake flour
3/4 teaspoon baking soda
1/2 teaspoon salt
1 cup granulated sugar
1/4 cup butter, softened

$1^1/_2$ cups mashed banana
(about 3 medium ripe bananas**)
2 large eggs
1/2 cup low-fat buttermilk
1/2 cup semisweet chocolate chips

1. Preheat the oven to 350°F. Grease and flour a 9 × 4 inch loaf pan and set it aside.
2. Sift together the flour, baking soda, and salt; set it aside.
3. Place sugar and butter in a medium-sized bowl; beat with a mixer at medium speed until well blended (about 1 minute).
4. Add banana, eggs, and buttermilk; beat until well blended. Add flour mixture; mix at low speed until the flour is moistened and not visible in the batter.
5. In a small microwavable safe bowl, place the chocolate chips and microwave for 15 seconds at HIGH. Remove the bowl from the oven. Stir the chocolate chips. Place back in the oven and microwave for another 15 seconds. Stir; repeat this procedure until the chocolate is melted and smooth.
6. Add the melted chocolate to **1 cup of the batter**, stirring to combine.
7. Spoon chocolate batter alternately with plain batter into the prepared loaf pan. Swirl together using a knife.
8. Bake for 1 hour and 15 minutes or until a wooden toothpick comes out clean. Cool for 10 minutes and then remove from pan onto a wire rack to cool.

**Ripe banana has a yellow skin with brown flecks.

D. PRUNE–RAISIN BROWNIES

1/4 cup seedless raisins
3/4 cup pitted prunes
$1^1/_2$ cups water, divided
$1^1/_2$ cups cake flour
$1^1/_3$ cups granulated sugar
1 cup unsweetened natural cocoa
1 teaspoon instant coffee
2 teaspoons baking powder

1/2 teaspoon baking soda
1/2 teaspoon salt
1 large egg
2 large egg whites
2 tablespoons vegetable oil
2 teaspoons vanilla extract
3 tablespoons chopped walnuts or pecans

1. Preheat the oven at 350°F. Coat a 9 × 13 inch pan with cooking spray.
2. Combine the prunes and raisins with 1 cup of water in a saucepan and bring to a boil; lower the heat and cover the pan, and cook the fruit for 5 minutes. Remove from the heat and stir in the remaining 1/2 cup of water. Transfer the mixture to a food processor or blender and puree the mixture; set aside.
3. Sift together the flour, sugar, cocoa, instant coffee, baking powder, baking soda, and salt into a large mixing bowl. Make a well in the center and add the egg and the egg whites, oil, vanilla, and pureed fruit. With a whisk or an electric mixer on low speed, blend the ingredients just until moistened. **Overmixing will produce tough and rubbery brownies.**
4. Spread the batter evenly in the prepared pan and sprinkle on top of the batter the chopped walnuts. Bake for about 30–35 minutes, or until a tester in the center comes out clean; do not over bake. Cool on a wire rack; then cut into 24 brownies.

E. APPLE-GINGERBREAD COBBLER

4 medium-sized cooking apples (such as
   Rome Beauty or Granny Smith), peeled and sliced thin
1 cup water
1/2 cup firmly packed light brown sugar
1 tablespoon lemon juice
1/4 teaspoon ground cinnamon
2 teaspoons cornstarch
1 tablespoon water
1/2 cup low-fat buttermilk
1/4 cup granulated sugar

1/4 cup unsulfured, mild flavored molasses
2 tablespoons vegetable oil
1 large egg
1/2 cup cake flour
1/2 cup whole wheat flour
1/2 teaspoon baking soda
1/2 teaspoon baking powder
1/4 teaspoon salt
1 teaspoon ground ginger
1/2 teaspoon ground cinnamon

1. Preheat the oven to 350°F.
2. Combine the first five ingredients in a medium saucepan; stir well.
3. Bring the mixture to a boil; cover and reduce heat; simmer for 10 minutes or until apple slices are tender.
4. Combine cornstarch and 1 tablespoon water; add to apple mixture and stir well. Bring to a boil.
5. Pour the apple mixture into a greased 2-quart baking dish.
6. Combine buttermilk, 1/4 cup sugar, molasses, oil, and egg in a small mixing bowl; beat at medium speed with an electric hand mixer until mixture is smooth.
7. Sift together the cake flour, whole wheat flour, baking powder, baking soda, salt, ginger, and cinnamon.
8. Add the dry ingredients to the liquid ingredients, and beat until blended.
9. Pour the batter over the apple mixture that is in the casserole dish.
10. Bake for 35 minutes or until a toothpick comes out clean when placed in the gingerbread. Serve warm.

F. BAKED APPLE CRUMBLE

Apple Layer

2 cups each: Golden Delicious, Rome Beauty, and
   Granny Smith, peeled and sliced

2 tablespoons orange juice

Topping**

1/2 cup light brown sugar, firmly packed
1/3 cup all-purpose flour

3/4 teaspoon cinnamon
4 tablespoons butter or margarine

1. Heat oven to 375°F. Spray a 13 × 9 inch oblong baking dish with vegetable spray.
2. Toss sliced apples with orange juice and place the apples in an even layer into the pan.
3. Make topping: combine brown sugar, flour, and cinnamon. Mix in butter until crumbly. Sprinkle on top of the apples.
4. Bake for 35 minutes or until apples are tender. Test with a fork to determine the tenderness; bake longer if necessary. Cool slightly. Serve warm as is or topped with a scoop of vanilla frozen yogurt.

**If you like more topping, you could double the recipe.

G. ORANGE POACHED PEARS

1/2 cup sugar
1/4 cup water
1/4 cup orange juice
6 whole cloves

1 tablespoon lemon juice
3 pears (bartlett or anjou)
Twist of orange rind (optional)

1. Combine sugar, water, orange juice, cloves, and lemon juice in a 2-quart saucepan; bring to a boil over medium heat, until sugar dissolves. Boil gently, cover for 5 minutes.
2. Peel pears and cut into chunk-sized pieces.
3. Add pears to the hot syrup. Cover, and simmer for 15 minutes.

4. Transfer pears and syrup to a medium-sized bowl; remove cloves; cover and chill thoroughly.
5. Spoon pears and syrup into dessert dishes; top with a twist of orange rind if desired.

H. BROILED GRAPEFRUIT

1 medium-sized grapefruit, cut in half   1 tablespoon butter or margarine, melted
4 teaspoons light brown sugar

1. Preheat broiler.
2. Cut around grapefruit sections to loosen, then sprinkle each grapefruit half with 2 teaspoons brown sugar and drizzle with 1/2 tablespoon melted butter.
3. Broil 5 inches from heat for 5–7 minutes until golden and serve hot.

Variations

1. *Honey-Broiled Grapefruit*: Substitute 1 tablespoon honey for each teaspoon of brown sugar and proceed as directed.
2. *Broiled Oranges, Peaches, Apricots, Pears, or Pineapples*: Halved oranges, peaches, apricots, and pineapple rings can be broiled. Sprinkle cut sides with brown sugar and butter, and broil for 3–5 minutes until delicately browned.

GENERAL QUESTIONS

1. Why should fruit be boiled gently?

2. What principle was involved when the dried fruit were cooked in the water for the Prune Brownies?

3. Why are dried fruit sulfured? Why should this be declared on the label of ingredients?

4. What does the term *tenderize* mean on the label of dried fruit?

5. What are the principal nutritional contributions of fruits?

6. What is the difference between a drupe and a pome fruit?

7. Why would the pears maintain their shape in the poaching recipe?

8. a. What pigment is common to berries, such as blueberries, raspberries, and blackberries?

   b. How nutritionally adequate are berries?

   c. What precautions should be followed when cooking fruit containing the pigment in (a)?

9. What gas is involved in the ripening process of fruit?

LABORATORY 8

Vegetables

LABORATORY 8
Vegetables

Vegetables provide versatility and variety to a meal. When preparing vegetables the student must preserve the color, texture, and nutrients that are inherent to a particular vegetable. This laboratory exercise will introduce the student to the proper techniques of preparing vegetables as well as the use of vegetables for different parts of a meal.

VOCABULARY

anthocyanin	carotenoids	flavonoids
anthoxanthin	chlorophyll	pheophytin
betalain	chlorophyllin	

OBJECTIVES

1. To learn how to recognize quality attributes in selecting fresh vegetables.
2. To learn how to maintain color, texture, flavor, and nutrients in the proper preparation of vegetables.
3. To recognize the versatility that vegetables contribute to the diet and to the meal.

PRINCIPLES

1. Vegetables come from different parts of a plant: leaf, stem, fruit, flower, root, tuber, bulb, and seed.
2. Vegetables can be cooked in a variety of ways: boiling, broiling, baking, steaming, frying, oven frying, stir-frying (or panning), microwaving, or pressure cooking.
3. Stir-frying and steaming are considered as the *Conservatory Methods* in cooking vegetables since these methods conserve the **texture**, **color**, and **nutrients** of the vegetable.
4. The color pigments in vegetables are categorized into
   a. chlorophyll: green.
   b. carotenoids: yellow, orange, or red.
   c. anthocyanins: red, purple, or blue.
   d. anthoxanthins: white or colorless.
   e. betalains: purplish red.
5. Anthocyanins and anthoxanthins have many similarities in their chemical structure; therefore, they are classified as flavonoid pigments.
6. Vegetables, as well as fruits, that contain the anthocyanin pigment are sources of antioxidants and phytochemicals which have strong nutritive connotations.
7. In an alkaline medium, such as provided by adding baking soda to the cooking water:
   a. green color (chlorophyll) will intensify: chlorophyllin will form
   b. very little effect on carotenoid pigment
   c. anthocyanins turn blue-green
   d. anthoxanthins turn yellow
   e. betalains turn brown in color
8. In an acid medium such as adding lemon juice, cream of tartar, or vinegar to the cooking water:
   a. chlorophyll (green) will turn olive green (pheophytin)
   b. no or little effect on carotenoids
   c. anthocyanins will retain their color (reddish)
   d. anthoxanthins are bleached or colorless
   e. betalains will maintain their color
9. The addition of an acid greatly retards the softening of the cellulose, making it very difficult to cook vegetables to the correct degree of doneness.
10. In an alkaline medium, the cellulose will soften very rapidly, leading quickly to a very mushy texture as well as the loss on vitamin C and thiamin (vitamin B_1).
11. Boil vegetables whole and unpeeled whenever possible. The smaller pieces, the greater will be the nutrient loss.

I. TO STUDY THE EFFECT OF HEAT AND pH ON PIGMENTS AND TEXTURE IN VEGETABLES

A. PROCEDURE

1. Place 1 cup water in each of four 1-quart saucepans.
2. Place the amount of alkaline (baking soda) or acid (cream of tartar) as indicated in Table 8.1.
3. Place a small amount of the vegetable in each saucepan.
4. Bring the water in each pan to a boil. Cover the pan and lower the heat so that the water is at a low boil. **It is imperative that the water is boiling in order to observe a change in color and texture**. Cook the vegetables as indicated in Table 8.1.
5. Place vegetable on a white plate.
6. Place cooking liquid into a small custard cup.
7. Record your observations in Table 8.1.

Table 8.1 TABLE FOR EVALUATING THE EFFECT OF HEAT AND pH ON COLOR AND TEXTURE OF VEGETABLES						
Pigment	Color	Vegetable	Baking Soda; 15 min	Cream of Tartar; 15 min	15 min	30 min
Chlorophyll	Green					
Betalains	Red–purple					
Anthoxanthins	White					
Carotenoids	Yellow or orange					
Anthocyanins	Red–blue					

QUESTIONS

1. What effect did the baking soda have on

   a. carotenoids?

   b. betalains?

   c. anthocyanins?

   d. anthoxanthins?

   e. chlorophyll?

    f.   texture of each vegetable?

    2.   What effect did the cream of tartar have on

        a.   carotenoids?

        b.   betalains?

        c.   anthocyanins?

        d.   anthoxanthins?

        e.   chlorophyll?

        f.   texture of each vegetable?

    3.   What effect did prolonged moist heat have on maintaining texture and pigmentation?

    4.   What recommendations would you give in cooking vegetables regarding maintenance of color, nutrients, and texture?

II. TO PREPARE VEGETABLES IN A VARIETY OF WAYS AND TO RECOGNIZE QUALITY PARAMETERS IN SELECTION AND PREPARATION

A. PREPARATION OF MEALY AND WAXY POTATOES BY CONVENTIONAL AND MICROWAVE METHODS

Potatoes of different varieties are available in various parts of the country. Basically, potatoes can be classified as mealy or waxy. Usually, potatoes that are round in shape will tend to be waxy (red potato), whereas the long flat ones tend to be starchy (Idaho or Russet). Mealy potatoes are preferred for mashing, baking, and frying. Waxy potatoes are a good choice for boiling or in recipes requiring the potato to hold its shape (salads, stews). There are also all-purpose potatoes that are a cross between a mealy and waxy potato (Yukon Gold is an example).

1. Baked Potatoes—Oven Method

1.   Preheat oven to 400°F.
2.   Select a baking potato, a red potato, and an all-purpose potato.

3. Place washed, dried potato on a shallow baking pan. Do not crowd potatoes on the baking pan. Prick several places with a fork.

4. Bake until potatoes can be pierced easily with a fork. Medium-sized potatoes in a 400°F oven may require 1 hour or more to bake. **Note:** If skins are to be eaten, a small amount of fat can be rubbed over the surface of the potato skin to give it a more tender skin on the baked potato.

5. As soon as potatoes are removed from the oven, the potatoes should be split open to allow the steam to escape and prevent the potatoes from becoming soggy.

6. Record observations in Table 8.2.

2. **Baked Potatoes—Microwave Method**

1. Select a baking potato, a red potato, and an all-purpose potato.

2. Scrub potatoes and dry thoroughly.

3. Pierce potatoes to allow steam to escape.

4. Arrange potatoes about 2 inches apart in a circle in the microwave (Fig. 8.1). (If potatoes are old, long, or dry, wrap each in waxed paper before placing in the microwave. For crisper skins, sprinkle dampened potatoes with salt.)

5. Microwave potatoes uncovered on HIGH (100%) until tender for 11–13 minutes. Let it stand for 5 minutes. (Potatoes hold their heat well; if microwaving a second vegetable be sure to cook the potatoes first.)

6. Record observations in Table 8.2.

FIG. 8.1: Arrange potatoes about 2 inches apart in a circle in microwave.

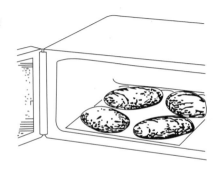

Table 8.2 TABLE FOR EVALUATION OF BAKED POTATOES				
Variety	Method	Appearance	Flavor	Moistness and Texture
Baking potato	Conventional			
Red potato	Conventional			
All-purpose potato	Conventional			
Baking potato	Microwaved			
Red potato	Microwaved			
All-purpose potato	Microwaved			

CHARACTERISTICS OF HIGH-QUALITY BAKED WHITE POTATOES

Appearance:	*White, opaque, or slightly translucent depending on variety.*
Texture:	*Cell structure is pliable, mealy, light.*
Moistness:	*Dry.*
Flavor:	*Bland, yet slightly sweet.*

B. COMPARISON OF TWO COOKING METHODS FOR BROCCOLI: STEAMING VERSUS MICROWAVE

1. Broccoli: Steaming Method

1. Place 1–2 inches of water in a 2-quart saucepan. Enough water must be used to produce steam for entire cooking time.
2. Place cleaned broccoli, cut into flowerets, into steamer basket; position over water. Tightly cover pan.
3. Bring water quickly to a boil over high heat. When water is boiling, reduce heat to keep water boiling slowly.
4. Cook broccoli for approximately 5–7 minutes. Test with a fork until tender.
5. Record observations in Table 8.3.

2. Broccoli: Microwave Method

1. Trim off large leaves; remove tough ends of lower stems; wash broccoli.
2. For spears, cut broccoli lengthwise into thin stalks. If stems are thicker than 1 inch, make lengthwise gashes in each stem.
3. Place 1/4 cup water and 1/8 teaspoon salt in baking dish, 12 × 7½ × 2 inches, or a 10-inch glass pie dish.
4. Arrange broccoli stalks in baking dish (tips in center) (Fig. 8.2) or arrange in circle in a 10-inch pie plate with tips in the center of the plate.
5. Cover with plastic wrap and microwave on HIGH (100%) for 4 minutes; rotate dish or pie plate 1/2 turn.
6. Microwave until tender, 3–5 minutes longer; drain.
7. Record observations in Table 8.3.

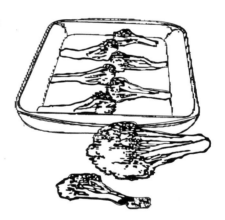

FIG. 8.2: Arrange broccoli stalks in baking dish (tips in center).

Table 8.3 TABLE FOR THE EVALUATION OF BROCCOLI			
Preparation Method	Color	Texture	Flavor
Steaming			
Microwaving			

Cheese Sauce for Broccoli (Optional)

1 cup milk
2 tablespoons butter or margarine
2 tablespoons all-purpose flour

1/4 teaspoon salt
1 cup mild Cheddar cheese, grated

1. Melt butter in a saucepan; whisk in flour and salt.
2. Remove the saucepan from the heat; whisk in milk.
3. Return the saucepan to the heat, and bring the mixture to a boil, stirring constantly. Boil for 1 minute.
4. Remove the saucepan from the heat. Add cheese. Stir until the cheese is melted and the cheese is blended into the sauce.
5. Place the hot cooked vegetable in a warm dish. Pour the hot sauce over the vegetable. **Use only sufficient sauce to cover; do not drown the vegetable in the sauce.**

QUESTIONS

1. What cooking method produced the best retention of color in the broccoli and why?

2. Which cooking method would help in retaining nutrient quality of the broccoli and why?

3. For potatoes, which method would you recommend: conventional or microwave?

4. Why is there a tendency to overcook potatoes in the microwave? What is meant by standing time, and how does it affect the cooking quality of the potato?

5. Why are the potatoes pricked with a fork before they are cooked in the conventional or microwave oven?

III. VEGETABLE PREPARATION

Vegetables can be used for different parts of a meal. This section illustrates the versatility of the vegetable and how it can service the meal not only as an accompaniment, but as a main course and as a dessert. Steaming and stir-frying (panning) have been cited as the Conservatory Methods for cooking vegetables. This section will also introduce "oven frying" and oven roasting vegetables in order to enhance their appeal and flavor.

A. ACCOMPANIMENT

1. Green Bean Casserole

1 pound green beans, cleaned and cooked	3 tablespoons skim milk
1 tablespoon butter or margarine	1/2 teaspoon Worcestershire sauce
2 tablespoons finely chopped onion	1/8 teaspoon red pepper
6 oz. fresh mushrooms, thinly sliced	1/2 cup shredded sharp Cheddar cheese
1 (14 oz.) can low-fat cream of mushroom soup	3/4 cup can fried onion rings
1/4 cup slivered almonds	

1. Clean string beans and remove the ends. Cut the string bean into thirds. In a large pot add water just one-third of the way. Bring the water to a boil and add **1 teaspoon salt**. Add the string beans and bring the water back to a boil. Cover and lower the heat. Cook the string beans for 8 minutes. Shock the cooked string beans by adding them to ice water to cool fast and retain their green color. Remove from ice water after 5 minutes and allow the string beans to drain while you make the sauce.
2. Preheat the oven to 350°F and grease a shallow 1½ quart baking dish.
3. In a 12-inch skillet melt butter over medium heat. Add onion and sauté for 5 minutes or until tender.

4. Add mushrooms and cook for 5 minutes longer or until soft.
5. In a large bowl, mix together the mushrooms/onion with the cooked string beans, soup, almonds, skim milk, Worcestershire sauce, and red pepper.
6. Spoon into baking dish and sprinkle with the Cheddar cheese. Bake uncovered for 20 minutes. Top with fried onion rings and bake for 5 minutes.
7. Remove from the oven and serve.

2. Healthy French Fries

1½ pounds large baking potatoes (Russet or Idaho)　　1 tablespoon olive oil
Chili powder　　Kosher salt to taste

1. Preheat oven to 425°F. Place a 10 × 15-inch jelly roll pan in the oven while it is pre-heating. **This is very important since if the potatoes are placed on a cold pan they will stick**.
2. Peel potatoes; cut potatoes lengthwise into 1/2 inch slices; then cut each slice into 1/2 inch thick strips. Place potatoes into a large bowl and toss with chili powder (**sprinkle liberally over the potatoes**). Add olive oil and toss again. Place potatoes in a single layer on the hot baking sheet.
3. Bake for 20 minutes; turn potatoes with a spatula and continue baking for 20 minutes longer, or until crisp and brown. Sprinkle with salt and serve hot.

3. Cauliflower Au Gratin

1 large cauliflower, divided into flowerets,　　1½ cups hot cheese sauce
　boiled and drained　　　(see Cheese Sauce for Broccoli)
1/2 teaspoon salt　　1/3 cup sharp Cheddar cheese, coarsely grated
1/8 teaspoon white pepper　　1 tablespoon butter or margarine, melted

1. Preheat broiler.
2. Arrange flowerets in an ungreased au gratin or shallow casserole.
3. Season cauliflower with salt and white pepper, and cover evenly with Cheese sauce.
4. Sprinkle with the grated cheese, and drizzle with the melted butter.
5. Broil 5 inches from heat for 3–4 minutes or until cheese melts and is dappled with brown.

4. Sweet–Sour Red Cabbage

1 tablespoon butter or margarine　　1 tablespoon Dijon mustard
1 medium onion, sliced　　3 tablespoons balsamic vinegar
1 teaspoon dry mustard　　1/2 cup water
2 tart medium apples, sliced　　1/2 teaspoon salt or according to taste
2 tablespoons light brown sugar　　1/4 teaspoon ground black pepper
1 pound red cabbage, finely shredded

1. Heat butter in a large skillet. Add onion and dry mustard. Cook until onion is tender.
2. Add apple slices and brown sugar; cook for 5 minutes.
3. Add cabbage, Dijon mustard, vinegar, water, and seasonings.
4. Bring mixture to a boil; lower heat and cover.
5. Cook over low heat for approximately 45 minutes or until cabbage is softened. Stir mixture often during the cooking period.

5. Stuffed Tomatoes

4 large, firm ripe tomatoes　　2 tablespoons parsley, finely chopped
1/2 teaspoon salt　　3/4 teaspoon salt
1/8 teaspoon black pepper　　1/8 teaspoon ground black pepper

Stuffing:

1 tablespoon butter or margarine
1/4 cup yellow onion, finely chopped
1 clove garlic, finely chopped
2 cups **soft bread crumbs**

1 cup tomato pulp (saved from tomato centers), chopped
1/4 cup grated Parmesan cheese
1/2 teaspoon dried oregano

1. Heat oven to 375°F.
2. Cut a thin slice from the top of each tomato.
3. Using a teaspoon, scoop out pulp and seeds; coarsely chop pulp and reserve 1 cup (use rest in soup or stew).
4. Sprinkle the inside of the tomatoes with the salt and pepper.
5. Melt butter in a small skillet over moderate heat; add onion and garlic; sauté for 8–10 minutes until golden.
6. Mix onion and garlic with remaining stuffing ingredients except the Parmesan cheese and oregano. Stuff tomatoes, then sprinkle top of stuffing with Parmesan cheese and oregano. Stand tomatoes in a buttered shallow casserole dish and bake uncovered, for 1/2 hour or until tender.

6. Ratatouille

1 medium eggplant, peeled, cut into 1/4 inch cubes
2 medium zucchini, sliced
2 teaspoons salt
1/4 cup olive oil
2 medium yellow onions, peeled and sliced thin
1 medium green bell pepper, cut into julienne strips
1 medium red pepper, cut into julienne strips

2 garlic cloves, peeled and crushed
2 medium tomatoes, coarsely chopped
1/4 teaspoon ground black pepper
1 teaspoon garlic salt
1/2 teaspoon dried Italian seasoning
2 tablespoons minced parsley
2 cups frozen whole kernel corn, thawed

1. Cut eggplant into 1/4 inch cubes and stir-fry, about 1/4 at a time, 3–5 minutes in 2–3 tablespoons olive oil over moderately high heat. Drain on paper towel (eggplant soaks up oil, but do not use more than 2–3 tablespoons for each batch or the ratatouille will be greasy).
2. Brown zucchini in 2 tablespoons of olive oil; drain.
3. Stir-fry onions, garlic, and green and red peppers in remaining olive oil over moderate heat for 10 minutes; lay tomatoes on top; add garlic salt, black pepper, parsley, and Italian seasoning; cover and simmer for 8–10 minutes; add corn.
4. Uncover and simmer for 10 minutes longer.
5. In an ungreased shallow 2½ quart flameproof casserole, build up alternate layers as follows: onion/pepper mixture, eggplant/zucchini (sprinkle with salt), onion/pepper mixture, eggplant/zucchini (sprinkle with salt), and onion/pepper mixture.
6. Simmer, covered over low heat (do not stir) for 30 minutes; uncover, simmer for 20 minutes longer.

7. Oven Roasted Vegetables

2 potatoes, cut into 1 inch chucks or cut in half lengthwise and sliced crosswise into 1/2 inch thick semicircles
2 sweet potatoes, peeled and cut into 1 inch chucks or sliced as above
2 carrots, peeled and cut into 1 inch chucks or cut on the diagonal into 1/2 inch thick slices
1 large onion, cut into wedges
1 medium zucchini, cut into 1 inch chunks or cut on the diagonal into 1/2 inch thick slices
1 yellow bell pepper, cut into 1 inch chunks
1 red bell pepper, cut into 1 inch chunks

Italian Dressing

4 teaspoons olive oil
1/4 cup fresh lemon juice
6 garlic cloves, finely minced
3 tablespoons fresh minced rosemary
(or 1½ teaspoons dried ground rosemary)

1 tablespoon fresh oregano
  (or 1 teaspoon dried oregano)
1 teaspoon salt

1. Preheat the oven to 425°F.
2. Parboil the potatoes, sweet potatoes, and carrots in boiling water to cover for 2 minutes. Drain well. In a large mixing bowl, combine the parboiled vegetables with the onions, zucchini, and bell peppers.
3. To make the dressing, whisk all the ingredients together in a small bowl.
4. Toss the vegetables well with the dressing. Place the vegetables in a single layer on a large unoiled baking tray and bake stirring every 15 minutes, until all the vegetables are tender, about 45 minutes.
5. Serve warm. These vegetables are also good leftover and served at room temperature.

8. Baked Spaghetti Squash

1 (2 pound) spaghetti squash, halved and seeded
1/4 teaspoon salt
Dash black pepper

2 tablespoons butter or margarine
2 cups prepared spaghetti sauce, heated
Grated Parmesan cheese

1. Cut spaghetti squash in half, remove seeds, and place in a large pot of boiling water.
2. Boil for 5 minutes.
3. Preheat the oven to 375°F. Place each piece of squash, hollow side up, on a square piece of foil large enough to wrap it.
4. Sprinkle with salt and pepper and dot with butter. **Wrap tightly**.
5. Place wrapped squash on a baking sheet and bake for 45 minutes or until tender.
6. Unwrap and then, with a fork, scrape flesh into strands (Fig. 8.3). Toss with spaghetti sauce and grated Parmesan cheese and place back into shell.

FIG. 8.3: Scrape flesh into strands with a fork.

B. MAIN DISH

1. Minestrone Soup

1 can (10 oz.) red kidney beans
1 large yellow onion, peeled and minced
1 clove garlic, peeled and minced
3 slices turkey bacon
1 tablespoon vegetable oil
1½ quarts beef broth or 1/2 and 1/2 mixture of
  water and beef broth
2 medium-sized carrots, peeled and diced
1 medium potato, peeled and diced
1 cup finely shredded cabbage

1/4 cup celery, chopped
2½ tablespoons tomato paste
1/2 teaspoon dried basil
1/4 teaspoon dried oregano
1/8 teaspoon dried thyme leaves
1 teaspoon salt
1/8 teaspoon ground black pepper
1/2 tablespoon parsley, minced
1/2 cup dry ditalini or elbow macaroni
2½ tablespoons Parmesan cheese

1. Stir-fry onion, garlic, and turkey bacon in oil in a large heavy kettle (5–8 minutes) over moderate heat until onion is pale golden.
2. Add broth and all the remaining ingredients except parsley, pasta, and Parmesan cheese; cover; simmer for 1 hour; stirring occasionally.
3. Add parsley and pasta; cover and simmer for 15–20 minutes longer until pasta is tender. Stir frequently.
4. Taste for salt and adjust as needed. Stir Parmesan cheese into the soup or if you prefer, pass it separately.

2. Fettuccine Primavera

1 package (16 oz.) dried fettuccine

Vegetables
1 tablespoon butter or margarine
2 tablespoons olive oil
1 clove garlic, split
1 zucchini, sliced 1/4 inch thick
1/2 pound broccoli, cut into $1\frac{1}{2}$ inch flowerets
1/2 red pepper, cut into 1/4 inch strips
1/2 pound whole fresh snow pea pods, ends trimmed
4 oz. cremini mushrooms, sliced

Sauce
2 tablespoons butter or margarine
1 cup 2% milk
1 cup Parmesan cheese, grated
1/4 teaspoon salt
Dash ground black pepper
2 tomatoes, peeled and chopped

1. Start cooking fettuccine as package label directs.
2. **Prepare vegetables:**
   a. In 1 tablespoon hot butter and oil toss garlic, zucchini, broccoli, and red pepper; stir-fry for 5 minutes or until vegetables are just crisp.
   b. Add pea pods and mushrooms; cook for 1 minute.
   c. Cook vegetables, covered, for 1–2 minutes.
   d. Do not overcook. Discard garlic.
3. **Make sauce:**
   a. In a medium saucepan, heat butter and 2% milk until butter is melted.
   b. Remove from heat. Add 3/4 cup cheese, salt, and black pepper; mix well.
4. **Drain fettuccine:**
   a. Toss fettuccine with sauce; turn out onto a heated serving platter.
   b. Place vegetables on top.
   c. Arrange tomatoes around edge.
   d. Sprinkle with 1/4 cup Parmesan cheese. Toss at table just before serving.

3. Eggplant Parmigiana

6 tablespoons all-purpose flour
1/4 cup toasted wheat germ
3 tablespoons sesame seeds
1/4 teaspoon salt
2 large egg whites
1 tablespoon water
1 large eggplant ($1\frac{1}{2}$–2 pounds), sliced into 1/4 inch thick slices
2 tablespoons olive oil

Herb Tomato Sauce
1 teaspoon olive oil
1/2 cup chopped onion
2 cans (8 oz. each) tomato sauce
1/4 teaspoon dried basil
1/4 teaspoon dried oregano, crumbled
1/4 teaspoon salt
1/4 teaspoon ground black pepper
1 teaspoon chopped parsley
$1\frac{1}{2}$ cups shredded part-skim mozzarella cheese

1. Prepare sauce and let simmer while fixing the eggplant.
   a. In a medium-sized heavy saucepan, heat the olive oil over moderate heat.
   b. Add the onion and sauté, stirring occasionally, for 5 minutes or until softened.
   c. Lower the heat and add tomato sauce, dried basil and oregano, salt, pepper and bring to a simmer.

    d.   Cook covered, stirring occasionally, for 15 minutes.

    e.   Stir in the parsley.

2. Preheat oven to 450°F.
3. Lightly grease a baking sheet.
4. On a large plate, combine the flour, wheat germ, sesame seeds, and salt.
5. In a shallow bowl, stir together the egg whites and water.
6. Dip the eggplant slices first in the egg white mixture, then the flour mixture to coat.
7. Arrange coated eggplant slices on the baking sheet; drizzle each slice with the oil and bake for 10 minutes.
8. Lower the oven temperature to 400°F and bake for 10 minutes longer.
9. Turn the eggplant slices over and bake for another 10 minutes or until crisp, golden, and tender.
10. Place eggplant in a lightly greased 9 × 11 × 2 inch pan.
11. Spread sauce over eggplant, then sprinkle with $1\frac{1}{2}$ cups shredded mozzarella cheese.
12. Bake for 15–20 minutes until cheese is melted and casserole is bubbly.

<u>Alternate Coating</u>

$1\frac{1}{2}$ cups bread crumbs	1 large clove garlic, minced or pressed
1/3 cup grated Parmesan cheese	1 teaspoon dried basil

1. Combine together the bread crumbs, Parmesan cheese, garlic, and basil in a shallow bowl.
2. Follow Steps **5–12 in the above directions.**

4. <u>Zucchini Lasagna</u>

1 tablespoon olive oil	2 large egg whites, lightly beaten and divided
1 medium-sized onion, chopped (1 cup)	1/4 cup plain dry bread crumbs
2 teaspoons minced garlic	1/8 teaspoon ground black pepper, or to taste
1 cup sliced white mushrooms	1/2 teaspoon garlic salt
4 cans (8 oz. each) tomato sauce	$1\frac{1}{2}$ cups part-skim ricotta cheese
1 tablespoon dried basil, crumbled	1/4 cup minced leaf parsley
5 medium-sized zucchini ($1\frac{3}{3}$ pounds), sliced lengthwise, 1/4 inch thick	1/4 cup Romano or Parmesan grated cheese
	3/4 cup shredded part-skim mozzarella cheese

1. In a large heavy saucepan heat the olive oil over moderate heat.
2. Add the onion, garlic, and mushrooms and sauté for 5 minutes or until the onion is soft.
3. Add the tomato sauce and the basil and bring to a boil.
4. Lower the heat and simmer the sauce, stirring occasionally for 20 minutes or until the sauce has thickened and reduced to about 3 cups.
5. Remove the sauce from the heat and set aside.
6. Preheat the oven to 425°F.
7. Line a baking sheet with aluminum foil.
8. Place the zucchini slices in a single layer on the baking sheet. Brush the topsides with the egg white, then sprinkle the bread crumbs, garlic salt, and pepper evenly over the slices.
9. Bake for 20 minutes or until zucchini is golden and crisp tender when pierced with a knife.
10. Remove from the oven and set aside until the zucchini is cool enough to handle.
11. Reduce oven temperature to 350°F.
12. In a medium-sized bowl combine the ricotta, egg white, parsley, and the 3 tablespoons of the Romano cheese.
13. Grease an 11 × 7 × 2 inch baking dish.
14. Spread 1 cup of the tomato–mushroom sauce evenly on the bottom of the baking dish.
15. Arrange half of the zucchini on top of the sauce in a single layer, trimming the slices to fit if necessary.
16. Spread the ricotta–Romano mixture over the zucchini.
17. Sprinkle 1/4 cup of the mozzarella cheese over the top; spoon 3/4 cup sauce over the mozzarella.
18. Place the remaining zucchini in a single layer on top again, trimming again.
19. Spread the remaining $1\frac{1}{4}$ cups sauce evenly over the zucchini layer, then scatter the remaining mozzarella cheese and 1 tablespoon of Romano cheese over the sauce.
20. Place the dish on a baking sheet in case the casserole bubbles over.

21. Bake for 35–45 minutes or until the cheese bubbles and the lasagna is hot in the center.
22. Cover the pan loosely with foil if the top browns too quickly.
23. Remove the lasagna from the oven and **let it stand for 15 minutes** before cutting.

5. Broccoli Lasagna with Roasted Red Pepper Sauce

8 cups broccoli flowerets
12 lasagna noodles
15 oz. container part-skim ricotta cheese
1/4 cup grated Parmesan cheese + 3 tablespoons
1 large egg
2 tablespoons minced parsley
1 (12 oz.) jar roasted red peppers, drained
4 tablespoons butter or margarine

1/4 cup all-purpose flour
2 cloves garlic, minced
3 cups skim milk
1/2 teaspoon salt
1/2 teaspoon dried basil
1/4 teaspoon ground black pepper
2 cups Monterey Jack cheese

1. Preheat oven to 425°F.
2. Place a steamer basket in a large 6-quart saucepan, add water. Bring to a boil. Meanwhile *trim and cut broccoli into flowerets*. Add broccoli to steamer basket; **sprinkle with some dried basil**; cover and reduce heat. Steam for 4–5 minutes or until crisp tender.
3. Cook 12 lasagna noodles until tender. Drain; spray with vegetable spray to prevent them from sticking.
4. *Filling*: Mix together in a small bowl the ricotta cheese, 1/4 cup parmesan cheese, egg, and mince parsley; set aside.
5. *Sauce*: Drain the 12 oz. jar of roasted peppers. Place peppers into a blender container and puree until almost smooth. In a skillet:
   a. Melt 4 tablespoons butter. Stir in 1/4 cup all-purpose flour, and 2 cloves garlic, minced.
   b. Cook and stir for 1 minute. Stir in gradually 3 cups of skim milk and the pureed peppers. Cook and stir until the mixture comes to a boil (**mixture will be thin**).
   c. Remove sauce from the heat and stir in 1/2 teaspoon salt, 1/2 teaspoon dried basil, and 1/4 teaspoon ground black pepper.
6. Coat a 3-quart rectangular baking dish with non-stick cooking spray.
7. Assemble dish:
   a. Spread 3/4 cup of the sauce in the dish.
   b. Arrange three lasagna noodles over sauce.
   c. Carefully spread 1/3 of the ricotta mixture over the noodles.
   d. Top with 1/3 of the broccoli.
   e. Sprinkle 1/2 cup of the Monterey Jack cheese over the broccoli.
   f. Repeat layers two more times, beginning with the sauce.
   g. Top with remaining noodles and sauce.
8. Bake covered for 20 minutes. Uncover and sprinkle with the remaining 1/2 cup of Monterey Jack cheese and 3 tablespoons Parmesan cheese.
9. Bake for 10 minutes or until heated through. Let it stand for 10 minutes before cutting to serve.

6. Moussaka

Meat Sauce	Cream Sauce	Salt
1/2 cup finely chopped onion	1 tablespoon flour	3 tablespoons melted margarine
3/4 pound ground beef or lamb	1/4 teaspoon salt	1/4 cup grated Parmesan cheese
1 garlic clove, crushed	Dash of ground black pepper	3/4 cup grated Cheddar cheese
1/2 teaspoon dried oregano leaves	1 cup skim milk	1 tablespoon plain bread crumbs
1/2 teaspoon dried basil leaves	1 large egg, beaten	
1/4 teaspoon ground cinnamon	1 large eggplant (1½ pounds),	
1/2 teaspoon salt	washed and dried	
Dash ground black pepper		
1 can (8 oz.) tomato sauce		

1. **Make meat sauce**:
   a. In a large skillet add chopped meat and brown over moderate heat; add onion and garlic and continue cooking, stirring occasionally until onion is soft.
   b. Add oregano, basil, cinnamon, salt, pepper, and tomato sauce; bring to a boil, stirring. Reduce heat; simmer uncovered for 15 minutes. Remove from heat.
2. **Broil eggplant**:
   a. Halve unpeeled eggplant lengthwise; cut crosswise into 1/2 inch thick slices. Place slices on a lightly greased cookie sheet.
   b. Brush eggplant slices with melted margarine; sprinkle with salt. Broil 4 inches away from the heat, 4 minutes on each side or until tender and browned.
3. **Make cream sauce**:
   a. In a medium saucepan stir together the flour, salt, and pepper. Gradually whisk in the milk.
   b. Bring mixture to a boil with constant stirring.
   c. In a small bowl beat the egg until it is thoroughly blended. Add a small amount of the hot cream sauce to the egg (**tempering**).
   d. Return the mixture to the saucepan; mix well. Remove the saucepan from the heat. Do not allow the mixture to boil or else it will curdle and scramble.
4. **Assemble the casserole**:
   a. Preheat oven to 350°F. In the bottom of a $9 \times 9 \times 2$ cubic inch pan layer half of the eggplant overlapping slightly.
   b. Sprinkle with 1 tablespoon of Parmesan and Cheddar cheeses.
   c. Stir bread crumbs into meat sauce; spoon evenly over eggplant in casserole.
   d. Sprinkle with 1 tablespoon each of Parmesan and Cheddar cheeses.
   e. Layer rest of eggplant; overlapping as before.
   f. Pour cream sauce over all. Sprinkle with any remaining cheese.
5. **Bake casserole for 35–40 minutes**.

7. Spaghetti Squash-and-Vegetable Gratin

(This recipe could serve as a main dish or as an accompaniment to a meal.)

3 cups cooked spaghetti squash
  (see **A-8**, **Baked Spaghetti Squash**)
1 tablespoon olive oil
1½ cups diced zucchini
1½ cups diced yellow squash
3 cups diced portabella mushrooms
2 cloves garlic, minced
1/2 teaspoon dried basil
3/4 cup (3 oz.) shredded part-skim
  mozzarella cheese, divided

1/4 cup chopped fresh parsley
1/2 teaspoon salt
1/4 teaspoon ground black pepper
1 (14.5 oz.) can stewed tomatoes, undrained and
  chopped

Cooking Spray
1/2 cup fresh bread crumbs

1. Preheat oven to 450°F.
2. Heat oil in a large skillet over medium heat. Add the zucchini, yellow squash, and mushrooms; sauté for 10 minutes or until vegetables are tender. Add garlic and cook mixture for 2 minutes to cook the garlic.
3. Remove mixture from heat. Add 1/4 cup mozzarella cheese, basil, parsley, salt, pepper, and tomatoes; set aside.
4. Combine 1/4 cup of mozzarella cheese and shredded spaghetti squash. Arrange cooked squash mixture in a large gratin dish or shallow 1½ quart baking dish coated with cooking spray. Spoon the tomato mixture over squash.
5. Combine the remainder of the shredded mozzarella cheese (1/4 cup) and bread crumbs; sprinkle over tomato mixture.
6. Bake at 450°F for 15 minutes or until bubbly.

C. DESSERT

1. Carrot Cupcakes with Lemon Cream Cheese Frosting

Cupcakes
2/3 cup granulated sugar
3 tablespoon vegetable oil
1 teaspoon vanilla
1 large egg
1 cup finely grated carrots
1 (8 oz.) can crushed pineapple, drained
1 cup all-purpose flour
1 teaspoon baking powder
1/4 teaspoon baking soda
1/8 teaspoon salt
3/4 teaspoon cinnamon
1/8 teaspoon nutmeg

Frosting
1/4 cup (2 oz.) low-fat cream cheese, softened
1 tablespoon butter or margarine, softened
$1\frac{1}{3}$ cup confectioners' sugar, sifted
1/2 teaspoon lemon extract
3–5 drops yellow food coloring
3 tablespoons sweetened flake coconut (optional)

1. Preheat oven to 350°F.
2. To prepare cupcakes: Combine first four ingredients (sugar, oil, vanilla, and egg) in a mixing bowl and with a mixer beat at medium speed until ingredients are well blended. Add carrots and pineapple, beat well.
3. Sift together flour and next five ingredients (baking powder, baking soda, salt, cinnamon, and nutmeg). Add to sugar mixture and combine ingredients just until blended and there is no visible flour present.
4. Spoon batter into 12 muffin tins lined with paper liners. Bake for 20 minutes or until a wooden toothpick comes out clean. Cool in pan for 5 minutes; remove cupcakes from the pan and place on wire rack to cool thoroughly before frosting.
5. To prepare the frosting: beat the cream cheese, butter, and lemon extract in a medium-sized bowl with a mixer at medium speed for 2–3 minutes. Gradually add confectioners' sugar (in two additions); beat until blended. Add yellow color gradually until desired color is reached.
6. Spread frosting over cupcakes, sprinkle with coconut.

QUESTIONS

1. Why should water be boiling in the pot when the vegetables are added?

2. In the Red Cabbage recipe why are apple and vinegar added to the cabbage during the initial cooking? What method is this known as?

3. If the chief aim is to preserve color, how should the following vegetables be prepared:

   a. beets?

   b. broccoli?

    c.   carrots?

    d.   cauliflower?

    e.   spinach?

4.   What is the green pigment found on the surface of potatoes?

5.   Give examples of vegetables that are good sources of:

    a.   fiber?

    b.   vitamin A?

    c.   vitamin C?

    d.   folic acid?

6.   What is the difference between a perishable and semi-perishable vegetable, and how should they be stored?

7.   What vegetables are considered aromatic vegetables, and how effective are they in adding flavor to prepared dishes?

LABORATORY 9

Salads

LABORATORY 9
Salads

Cold, clean, crisp, and dry are the descriptive words for salad preparation. A salad adds color and texture to a meal, therefore, the proper selection and combination of ingredients are important. The student will learn the proper care of salad ingredients, as well as the preparation of an attractive and palatable product that will enhance a meal.

VOCABULARY

accompaniment salad   dessert salad   main course salad
appetizer salad   garnish   salad greens

OBJECTIVES

1. To observe the proper care and handling of salad greens.
2. To learn how color plays an important role in creating a salad.
3. To learn how flavor and texture balance is important in a salad.

PRINCIPLES

1. Salad greens must be washed and dried thoroughly before use.
2. Salad greens include: Iceberg lettuce; Romaine Lettuce (cos); Bibb lettuce; Boston lettuce; green leafy lettuce; red leafy lettuce; spinach; arugula; and radicchio.
3. Lukewarm water will open up crinkly-type leaves, such as, spinach for thorough cleaning.
4. It is suggested that green leaves that are used for salads should be torn and not cut into bite-size pieces.
5. The size of the pieces of salad greens should be big enough to fit into the mouth, and not so small that they lose their identity.
6. Color and texture contrasts are important when creating a salad.
7. The dressing for the salad must be added right before the salad is to be served. If the dressing is added ahead of time, the greens will become wilted and soggy. A marinated salad, such as coleslaw, potato salad, contains ingredients that do not wilt when held with the dressing added ahead of time.
8. There are four types of salad:
   a. Appetizer: wets the appetite.
   b. Accompaniment salad: complements the meal (simple [few ingredients] for an elaborate meal; complex [variety of vegetables] for a simple meal).
   c. Main dish salad: contains a protein in the form of chicken, fish, cheese, eggs, etc.
   d. Dessert salad: served at the end of a meal.

I. APPETIZER SALAD

A. ANTIPASTO SALAD PLATTER

2 medium red peppers*   2 tablespoons olive oil
1/2 teaspoon salt   1 clove garlic, minced

*May use a combination of red and green peppers for color.

1. Preheat oven to 450°F.
2. Wash peppers; drain well.
3. Place peppers on a flat baking sheet; bake about 20 minutes, until skin becomes blistered and charred. With tongs turn peppers every 5 minutes.
4. Place cooked peppers in a large saucepan. Cover and let stand for 15 minutes.

5. Peel off charred skin with a sharp knife. Cut each pepper into fourths. Remove ribs and seeds; cut out any dark spots.
6. In a bowl combine oil, salt and garlic; add peppers and toss lightly to coat; refrigerate.

Deviled Eggs

4 eggs
1/2 teaspoon dry mustard

2 tablespoons mayonnaise
1/8 teaspoon salt

1. Place eggs in a small saucepan and cover with water. Bring water to a boil and then lower heat. Gently boil the eggs for 10 minutes. Remove from heat and immediately cool the eggs under cold running water for 5 minutes.
2. Peel the eggs and cut lengthwise. Scoop out the yolk; mash the yolk with a fork and add the mayonnaise, dry mustard, and salt; mix well.
3. Place prepared yolk mixture into the egg white shell; refrigerate.

Assembly

1/4 pound sliced Genoa salami
1/4 pound sliced ham

1/2 cup slice black olives

1. Roll up salami and ham; place alternately on plate.
2. Place peppers in the center of the plate. Place Deviled eggs on the plate between the salami and ham. Place olives on the plate.

II. DINNER ACCOMPANIMENT SALADS

A. GREEN BEAN SALAD

2 pounds of green string beans (or 1 pound green and 1 pound yellow string beans)
$1\frac{1}{2}$ teaspoons salt

1. Add water, about 2–3 inches, in a large saucepan.
2. While water is coming to a boil, clean string beans.
3. When the water starts to boil add the salt and string beans. Bring the water back to a boil; cover the saucepan and cook the string beans for 8–10 minutes.
4. Drain beans in a colander, and immediately plunge the sting beans into ice water to hold the color and stop the cooking process. Dry the beans on a towel.
5. Place string beans in a large bowl and add the dressing (below).

Dressing

3 tablespoons white or red wine vinegar
$1\frac{1}{2}$ teaspoons Dijon mustard
1 teaspoon minced garlic
1 tablespoon mayonnaise

3/4 teaspoon salt
1/4 teaspoon ground black pepper
1/3 cup virgin olive oil

1. In a small bowl whisk together the vinegar, mustard, garlic, mayonnaise, salt, and pepper.
2. While whisking, slowly add the olive oil.
3. Add dressing to the string beans.

Finishing the Salad

1/2 small red onion, cut into rings
1/4–1/3 cup chopped fresh basil

3 oz. fat-free feta cheese, crumbled

1. Add the above ingredients to the salad and toss again to distribute the ingredients.

B. TANGY CARROT SALAD

5 cups thinly sliced carrots
2 tablespoons olive oil
1/4 cup white wine vinegar

1 teaspoon dried basil leaves
1/4 teaspoon minced garlic
1/2 teaspoon salt

Dash of ground black pepper

1. Cook carrots in boiling water to cover about 7 minutes or until crisp tender; drain.
2. Combine cooled carrots and remaining ingredients in a bowl; toss gently.
3. Refrigerate until ready to serve.

C. TABBOULEH

1 cup bulgur wheat
$1\frac{1}{2}$ cups boiling water
1/4 cup freshly squeezed lemon juice
2 tablespoons olive oil
$2\frac{1}{2}$ teaspoons salt
1/2 cup minced scallions

1/2 cup chopped fresh mint
1/2 cup chopped flat leaf parsley
1 hothouse cucumber, unpeeled, seeded, and medium-diced
1 cup cherry tomatoes cut in half
1/2 teaspoon ground black pepper

1. Place the bulgur in a large bowl, pour in the boiling water, and add the lemon juice, olive oil, and $1\frac{1}{2}$ teaspoons salt. Stir; then allow to stand covered at room temperature for about an hour.
2. Add the scallions, mint, parsley, cucumber, tomatoes, 1 teaspoon salt, and the pepper; mix well.
3. Season to taste and serve, or cover and refrigerate.

The longer this salad stands, the better the flavor develops in it.

D. PANZANELLA SALAD

8 oz. French style bread, cut into 1 inch cubes
8 oz. frozen broccoli flowerets
2 roasted red peppers, cut into thin strips
$1\frac{1}{2}$ cups Roma tomatoes, diced (about $3\frac{1}{4}$ cups)
$1\frac{1}{2}$ oz. Provolone cheese, diced
2 tablespoons pitted olives, coarsely chopped

1/3 cup fresh basil leaves
2 tablespoons balsamic vinegar
1 clove garlic, peeled
1 tablespoon olive oil
1/2 teaspoon salt

1. Preheat oven to 400°F. Spread the bread cubes onto a baking sheet and bake the cubes for 8 minutes, or until lightly brown and crisp.
2. Cook frozen broccoli in microwave for 6 minutes.
3. In a large salad bowl, mix together the pepper strips along with the broccoli, Provolone cheese, olives, and 2 cups of the diced tomatoes.
4. Add the remaining $1\frac{1}{4}$ tomatoes, basil, vinegar, oil, salt, and garlic to the food processor and puree until smooth.
5. Add toasted bread cubes to the salad bowl along with the vegetables. Pour dressing over the salad, toss to combine, and serve.

E. SPINACH–MUSHROOM SALAD

2 tablespoons vegetable oil
2 tablespoons white wine vinegar or balsamic vinegar
1/4 teaspoon salt
Dash of ground black pepper
1 garlic clove, crushed

1 pound spinach, cleaned and torn into bite-size pieces (about 8 cups)
8 oz. fresh mushrooms, sliced (about 3 cups)
2 slices turkey bacon, crisply cooked and crumbled
1 hard-cooked egg, chopped

1. Shake oil, vinegar, salt, pepper, and garlic glove in a tightly covered container.
2. Add dressing to the spinach and mushrooms in a large bowl; toss.
3. Sprinkle the tossed salad with chopped hard cooked egg and crumbled bacon.

F. TOSSED GREEN SALAD WITH CHEESE–MUSTARD DRESSING

4 slices of white bread, cubed
2 large garlic cloves, cut lengthwise in half
2 tablespoons olive oil or vegetable oil

4 oz. fresh mushrooms, sliced (about 1$\frac{1}{2}$ cups)
1/2 head Boston lettuce, torn into bite-size pieces
1/2 small bunch Romaine, torn into bite-size pieces

Grated Parmesan Cheese

1. Prepare Cheese–Mustard Dressing (below).
2. Cut bread into cubes about 1 inch square. Cook and stir garlic in oil in an 8-inch skillet over medium heat until garlic is a light brown; remove garlic and discard.
3. Add bread to oil. Cook and stir until bread is golden brown and crusty; cool.
4. Shake Cheese–Mustard Dressing; add Dressing to mushrooms, Boston lettuce, and Romaine and toss. Sprinkle the tossed salad with the croutons and Parmesan grated cheese.

Cheese–Mustard Dressing

1/4 cup olive oil
2 tablespoons red wine vinegar
1 tablespoon + 1 teaspoon grated Parmesan
  cheese

4 teaspoons Dijon mustard
1/2 teaspoon salt
1/2 teaspoon of each: dried oregano, basil,
  and dill

1. Shake all ingredients in a tightly covered container.
2. Refrigerate at least for 1 hour.

G. MANDARIN SALAD WITH SWEET–SOUR DRESSING

1/4 cup sliced almonds
1 tablespoon + 1 teaspoon granulated sugar
1/4 head of lettuce, torn into bite-size pieces
1/4 bunch Romaine, torn into bite-size pieces

2 medium stalks of celery, chopped (about 1 cup)
2 green onions, chopped
1 can (11 oz.) mandarin orange segments, drained

1. Cook almonds and sugar over moderate heat in a small heavy skillet, stirring constantly, until sugar is melted and almonds are caramelized (coated with the melted sugar). Immediately pour the almonds onto a lightly greased flat plate. **Mixture is very hot**. **Do not touch it**. Allow almonds to cool. Break apart the almonds.
2. Prepare Sweet–Sour Dressing (below).
3. Place lettuce and Romaine in a salad bowl; add celery and onions.
4. Add orange segments. Add Sweet–Sour Dressing.
5. Toss until all ingredients are evenly coated; add almonds and toss lightly.

Sweet–Sour Dressing

2 tablespoons vegetable oil
1 tablespoon granulated sugar
Dash ground black pepper

2 tablespoons apple cider vinegar
1/2 teaspoon salt

1. Shake all ingredients in a tightly covered container.

H. TRICOLOR CABBAGE SLAW

2 cups green cabbage, finely shredded
2 cups red cabbage, finely shredded
1/2 small green pepper, chopped (about 1/4 cup)
1 large carrot, peeled and grated
1/3 cup white vinegar
1 tablespoon vegetable oil

2 tablespoons granulated sugar
1 teaspoon instant onion
1 teaspoon salt
1/2 teaspoon celery seed
1/2 teaspoon dry mustard
1/4 teaspoon ground black pepper

1. Mix all ingredients in a large bowl; cover and refrigerate for 3 hours. Just before serving, drain the salad.

I. CREAMY CUCUMBER SALAD

1/2 cup plain yogurt
1/2 teaspoon salt
1/4 teaspoon dried dill weed

1/8 teaspoon ground black pepper
2 medium cucumbers, thinly sliced
1 small red onion, thinly sliced and separated into rings

1. Mix together the yogurt, salt, dill, and black pepper. Pour over the cucumbers and red onions that are in a shallow dish; toss. Cover and refrigerate for at least 4 hours.

J. TOSSED SALAD WITH PECANS AND STRAWBERRIES AND POPPY SEED DRESSING

4 cups salad greens
  (combination of leaf lettuce, Romaine, spinach)

1/2 cup pecan halves, toasted*
1 cup fresh strawberries, sliced

Dressing

1/3 cup apple cider vinegar
2 tablespoons granulated sugar
1 teaspoon salt
2 tablespoons vegetable oil

1 tablespoon Dijon mustard
1 teaspoon poppy seeds
1/4 teaspoon ground black pepper

1. Wash and dry the salad greens.
2. Combine Dressing ingredients in a small bowl by whisking. Refrigerate until ready to serve.
3. Toss greens, pecans, and sliced strawberries with dressing. It is not necessary to use all the dressing.
4. Serve salad immediately.

*Toast pecans in a nonstick skillet over medium heat. Stir with a wooden spoon until the pecans are fragrant. Remove pecans immediately from skillet and onto a plate to cool.

K. PASTA SALAD

8 oz. multicolored pasta spirals
1 tablespoon vegetable oil
1½ cups broccoli flowerets

1/4 cup chopped red onion
2 oz. pepperoni, sliced
1 oz. ripe pitted olives, drained and sliced

Dressing

1 tablespoon Dijon mustard
1/4 cup red wine vinegar
Dash ground black pepper
1 garlic clove, finely chopped
2 tablespoons minced flat leaf parsley

1 tablespoon granulated sugar
1/2 teaspoon salt

1/4 cup olive oil
1/2 cup grated Parmesan cheese

1. Cook pasta according to the package directions; drain and chill in ice water. Drain; add 1 tablespoon vegetable oil; refrigerate.
2. Blanch broccoli; drain and chill in ice water; drain.
3. Mix broccoli with pasta, red onion, pepperoni, and olives.

4. In a small bowl whisk together Dijon mustard, vinegar, sugar, salt, black pepper, garlic, and minced parsley.
5. Slowly whisk in the olive oil.
6. Add dressing and Parmesan cheese to pasta mixture; toss ingredients; chill. May be served on a bed of lettuce leaves.

III. MAIN DISH SALADS

A. CHEF'S SALAD

1/2 cup 1/4 inch strips of meat, cooked (beef, smoked ham, or tongue)
1/2 cup 1/4 inch strips of chicken or turkey, cooked
1/2 cup 1/4 inch strips of Swiss cheese
1/2 cup green onions (with tops), chopped
1 medium head lettuce, torn into bite-size pieces

1 small bunch Romaine, torn into bite-size pieces
1 medium stalk celery, sliced (about 1/2 cup)
1/2 cup mayonnaise or salad dressing
1/4 cup French Dressing (see below)
2 hard-cooked eggs, sliced
2 tomatoes, cut into wedges pitted ripe olives

1. Reserve a few strips of meat, chicken, and cheese. Toss remaining meat, chicken, and cheese, the onions, lettuce, Romaine, and celery.
2. Mix mayonnaise and French Dressing; pour over lettuce and toss. Top with reserved meat, chicken and cheese strips, the eggs, tomatoes, and olives.

French Dressing

1/4 cup olive oil
2 tablespoons red wine vinegar
2 tablespoons freshly squeezed lemon juice

1/2 teaspoon salt
1/4 teaspoon dry mustard
1/4 teaspoon paprika

Shake all ingredients in a tightly covered container; refrigerate. Shake before serving.

B. TUNA–MACARONI SALAD

1 cup elbow or spiral macaroni, uncooked
1 cucumber, chopped
1/2 cup low-fat mayonnaise or salad dressing
1 tablespoon onion, finely chopped
2 tablespoons sweet relish

1 tablespoon lemon juice
1/2 teaspoon salt
1/4 teaspoon ground black pepper
2 cans (6 oz., each) tuna in water, drained
4 cups salad greens, torn into bite-size pieces

1. Cook macaroni as directed on the package; drain.
2. Rinse the cooked macaroni under cold running water; drain.
3. Mix cooked macaroni and remaining ingredients except the salad greens.
4. Cover and refrigerate for at least 1 hour. Spoon onto salad greens. Garnish with tomato wedges if desired.

C. CHICKEN SALAD CUPS

Chicken–Almond Salad

2 cups chicken or turkey, cooked, cut up
1/2 cup celery, chopped
1/2 cup almonds, slivered and toasted
1/3 cup mayonnaise or salad dressing

1 tablespoon lemon juice
1/4 teaspoon salt
1/4 teaspoon ground black pepper
1 jar (2 oz.) sliced pimentos, drained

1. Mix all ingredients; refrigerate until chilled, at least 1/2 hour.

Salad Cups

1/2 cup water	1/2 cup all-purpose flour
1/4 cup butter or margarine	
Dash of salt	
1/2 teaspoon poppy seeds	2 eggs

1. Heat oven to 400°F. Grease six medium muffin cups, $2^1/_2 \times 1^1/_2$ inches.
2. Heat water and butter to rolling boil in a 2-quart saucepan; stir in flour and salt.
3. Stir vigorously over low heat until mixture forms a ball, about 30 seconds; remove from heat. Cool slightly.
4. Beat in eggs and poppy seeds all at once; continue beating until smooth.
5. Spread 2 rounded tablespoons of dough in bottom and sides of each muffin cup. Bake until puffed and dry in center, about 30 minutes. Immediately remove from pan. Cool on a wire rack.
6. Just before serving, fill each cup with the chicken filling.

IV. DESSERT SALADS

A. WINTER FRUIT SALAD WITH LIME–GINGER DRESSING

2 medium apples, cut into 1/4 inch slices	1 grapefruit, pared and sliced
2 oranges, pared and sliced	2 cups seedless grapes

Salad greens

1. Prepare Lime–Ginger Dressing (below).
2. Dip apples in dressing.
3. Arrange apples, oranges, grapefruit, and grapes on salad greens.
4. Serve with Lime–Ginger Dressing.

Lime–Ginger Dressing

1/3 cup frozen limeade concentrate, thawed	1/2 teaspoon ground ginger or 1 tablespoon of finely
1/3 cup honey	chopped crystallized ginger
1 tablespoon vegetable oil	

1. Combine all ingredients in a small mixing bowl.
2. Beat with a hand mixer until smooth.

B. EASY FRUIT SALAD

1 cup seedless grapes	1 can ($8^1/_4$ oz.) pineapple chunks in syrup, chilled
1 can (11 oz.) mandarin orange segments,	and drained
chilled and drained	1 red delicious apple, diced
Salad greens	
Fruit Salad Yogurt Dressing (recipe below)	

1. Mix grapes, orange segments, pineapple chunks, and apple.
2. Spoon onto salad greens. Serve with Fruit Salad Yogurt Dressing.

Fruit Salad Yogurt Dressing

2/3 cup plain low fat or non-fat yogurt   1 tablespoon orange juice
1 tablespoon honey   1/8 teaspoon almond extract

1.   Mix all ingredients together.

QUESTIONS

1.   What are the general rules that should be observed in determining the size of the pieces of salad greens to be used when preparing a salad?

2.   Describe the care of cleaning salad greens in preparation of a salad. Prior to serving, how can the salad greens be held in order to maintain their crispness?

3.   What points should be observed in the creation of an attractive salad?

4.   When fragile ingredients are used in a salad, what precautions should be observed?

5.   When and why should the dressing be added to a salad?

6.   From the recipes in the unit, identify those that are classified as a marinated salad.

7.   Name, define, and give an example of the four different salads that are used in planning a menu.

8.   What causes the browning of fruit, such as apples and bananas, in a fruit salad, and what should be used in the dressing to ensure an attractive salad?

9.   What should be the nutritional contribution of a salad to the daily diet?

10.   What should one consider when selecting salad greens and other ingredients to supplement the nutritional quality of the dish?

LABORATORY 10

Fats and Emulsions

LABORATORY 10
Fats and Emulsions

Fats contribute flavor and tenderness to food, but serve as a transfer of heat when used as a medium for cooking. Fat and water are insoluble, but when a third agent (an emulsifier) is used, these two liquids are brought together and an emulsion is formed. This laboratory exercise will look at different types of fats, their stability, and the role that fat plays in transferring heat during preparation. The difference between a temporary and permanent emulsion will be demonstrated.

VOCABULARY

acrolein	hydrogenated	phospholipid
antioxidant	hydrolytic rancidity	polyunsaturated fatty acid
butter	immiscible	rancidity
dispersing medium	lard	smoke point
double bond	margarine	temporary emulsion
emulsifier	monounsaturated fatty acid	triglyceride
fatty acid	oxidation	winterized
free fatty acid	permanent emulsion	
glycerol	plastic	

OBJECTIVES

1. To show the effect of frying temperature on the quality of the cooked product.
2. To evaluate various fats according to color, flavor, and aroma.
3. To be familiar with the rancid quality of fat.
4. To emphasize different emulsions and the effect of an emulsifying agent on the formation and the stability of the emulsion formed.

PRINCIPLES

1. Fatty acids can be long- or short-chained; saturated or unsaturated.
2. The melting point of a fat is determined by
   a. the length of the chain.
   b. the degree of saturation.
3. Fats that are used for frying should have a high smoke point.
4. The smoke point used for frying is important and will affect the quality of the cooked food.
   a. Too low a frying temperature will cause absorption of the fat by the food.
   b. Too high a frying temperature will cause burning and undercooking of the food product.
5. During frying, fats come off the triglyceride molecule in the form of free fatty acids.
6. The free fatty acids join together (polymerization) and cause thickening of the frying fat and a sticky build-up on the fry basket. The free fatty acids also lower the smoke point of the fat.
7. Glycerol is also broken down by dehydration in the frying fat, thereby, forming acrolein which is irritating to the nose and eyes.
8. Fats can go rancid in two ways:
   a. hydrolytic rancidity which is caused by enzymatic activity.
   b. oxidative rancidity which is caused by the exposure of the fat to oxygen, light, moisture, and metals.
9. Antioxidants are used in fats and oils to prevent rancidity. Typical examples are: butylated hydroxyanisole (BHA), butylated hydroxytoluene (BHT), tertiary butyl hydroquinone (TBHQ), and propyl gallate (PG).
10. Fat and water are immiscible and when they come together they form a temporary emulsion. With the aid of an emulsifier a permanent emulsion is formed.
11. Mayonnaise and a cooked salad dressing are examples of a permanent emulsion. French dressing is an example of a temporary emulsion.

106

CAUTION

Fats and oils can be heated to very high temperatures and are easily ignited when hot. Be careful not to burn yourself or spill oil on a hot burner. In case of fire, smother the fire with a lid or baking soda. **DO NOT PUT WATER ON A FIRE.** Place the deep fat frying pan with the handle away from you and not near the edge of the counter to avoid spills and possible burns.

I. TO SHOW THE EFFECT OF FRYING TEMPERATURE ON FAT ABSORPTION DURING DEEP FAT FRYING

A. DOUGHNUT HOLES

$2\frac{1}{4}$ cups all-purpose flour	1/2 cup sugar
2 teaspoons baking powder	1 tablespoon vegetable oil
1/2 teaspoon salt	1/2 cup milk
Dash of nutmeg	1 egg
1/8 teaspoon cinnamon	4 cups vegetable oil for frying (approximate)

1. Place fat for frying in a pan 5–6 inches deep with a diameter 6–7 inches. There should be 2–3 inches of liquid fat in the pan. Do not use less than 2 inches or more than 3 inches of fat. Do not start heating the fat until the doughnut holes are prepared.
2. Sift together all dry ingredients in a large size mixing bowl.
3. In a small size bowl, blend together the egg, milk, and vegetable oil. Use a rotary beater for blending. The oil must blend thoroughly with the egg and the milk as the liquid mixture is added to the dry ingredients.
4. Add the liquid ingredients to the dry ingredients and stir 50 strokes. Scrape down the sides of the bowl and clean the spoon with a rubber spatula midway in the stirring period. Place dough on a lightly floured surface.
5. Roll dough to a 1/2 inch thickness (use guides if possible to get uniform thickness). Cut dough with a doughnut hole cutter.
6. Weigh two doughnut holes and record weight:_____. Fry at 325°F. Drain and reweigh; record weight:_____.
7. Cook the remaining doughnut holes (except two) for approximately 3 minutes at 365°F. It will be necessary to turn the holes to get even browning during frying.
8. Use a slotted spoon to remove the cooked doughnut holes from the fat. Place the doughnut holes on several layers of paper toweling to drain off excess fat.
9. Weigh two doughnut holes cooked at 365°F and record weight:_____.
10. Raise the temperature of the hot fat to 390°F and cook the two remaining doughnut holes.
11. Weigh the two doughnut holes and record weight:_____.
12. Calculate the percentage of fat absorption using the following formula:

$$\frac{\text{cooked weight} - \text{precooked weight}}{\text{precooked weight}} \times 100$$

13. For evaluation **do not** roll the cooked doughnut hole in sugar. Doughnut holes not used for evaluation may be rolled in sugar.
14. Evaluation: Cut doughnut holes in half. Examine for penetration of oil used in frying. The absorbed fat will form a ring inside the doughnut hole. Record observations in Table 10.1.

Table 10.1 TABLE FOR EVALUATION OF DOUGHNUT HOLES				
Frying Temperature	Original Weight (g)	Fry Time (min)	Drained Weight (g)	Penetration of Fat
325°F				
365°F				
390°F				

QUESTIONS

1. What effect will

   a. a low frying temperature have on the quality of the food product?

   b. a high frying temperature have on the quality of the food product?

2. Describe how the composition of a food item will affect fat absorption during the frying period.

3. a. What was the color of the frying oil at the beginning and end of the frying period?

   b. What caused the color change?

   c. What happens to the smoke point of the fat with continuous use?

   d. What factors affect the smoke point of a fat?

4. Describe the desirable characteristics of fried foods.

5. What is the irritating ingredient in the smoke from frying fat? What causes it?

6. Suggest fats that are good for frying.

7. What are the structural differences between a saturated fatty acid and an unsaturated fatty acid? Which fatty acid is subject to trans-fat formation?

II. TO EVALUATE FATS ACCORDING TO COLOR, FLAVOR, AROMA, AND TO BE ABLE TO DETECT RANCIDITY IN A FAT

A. IDENTIFICATION OF FATS AND OILS

A series of fats and oils will be displayed for evaluation of aroma, color, and flavor. Both fresh and rancid varieties will be presented for identification. List the fats in Table 10.2 and record observations.

Table 10.2 TABLE FOR EVALUATION OF FATS AND OILS				
Type of Fat	Source	Color	Odor	Flavor

QUESTIONS

1. What causes a fat to become rancid?

2. Which fat would be more stable and why: a fat with a high saturated fatty acid content or a fat with a high unsaturated fatty acid content?

3. What steps must be taken to prevent rancidity in a fat?

4. a. What is an antioxidant? Identify those that are used in commercially processed fats and oils.

   b. What natural antioxidant is found in common vegetable oils?

5. What is a synergist? Describe its action in a fat or oil.

6. Name the types of lard and how they are processed.

7. What is the difference between butter and margarine?

8. What is meant by a refined vegetable oil?

9. What is the difference between virgin olive oil and light olive oil?

10. What are the nutritional consequences in consuming olive oil, soybean oil, and butter? Which one(s) should be incorporated into the daily diet and why?

III. TO BECOME FAMILIAR WITH VARIOUS EMULSIONS AND THE EFFECT THAT THE EMULSIFIER HAS ON STABILIZING THE EMULSION

A. FRENCH DRESSING (TEMPORARY EMULSION)

1 teaspoon granulated sugar	Dash of cayenne pepper
1/2 teaspoon salt	2 tablespoons lemon juice
1/2 teaspoon dry mustard	2 tablespoon red wine vinegar
1/2 teaspoon paprika	1/2 cup olive oil

1. Place all ingredients in a covered jar or container.
2. Shake well before using.
3. Record observations in Table 10.3.

B. MAYONNAISE (PERMANENT EMULSION)

2 large egg yolks	1/8 teaspoon paprika
1/2 teaspoon salt	1/4 teaspoon dry mustard
Dash of cayenne pepper	
1 tablespoon cider vinegar	1 tablespoon lemon juice
1 cup salad oil	

1. Place egg yolks in a **small deep bowl** (1 quart). Add salt, paprika, mustard, and cayenne pepper, and blend well with the egg yolks.
2. Add vinegar and mix well.
3. Add salad oil, a few droplets at a time, beating with hand mixer until a 1/2 cup of oil has been added.
4. Beat in the lemon juice.
5. Add the balance of the salad oil (1/2 cup) 1 tablespoon at a time; beating well after each addition.
6. Record observations in Table 10.3.

NOTE: Due to the risk of raw eggs containing *Salmonella*, mayonnaise should not be tasted.

C. BLENDER MAYONNAISE (PERMANENT EMULSION)

1 egg	1/2 teaspoon salt
1/2 teaspoon dry mustard	2 tablespoon vinegar (white or cider)
1/2 teaspoon granulated sugar	1 cup salad oil

1. Put egg, dry mustard, sugar, salt, vinegar, and 1/4 cup oil into the blender container.
2. Cover and process at **Blend** for 30 seconds.
3. Remove feeder cap and pour in the remaining oil in a slow and steady stream. (If necessary, STOP BLENDER, use rubber spatula to keep mixture around processing blades. Cover and continue to process.)

110

4. Store covered in the refrigerator for up to 1 week.
5. Record observations in Table 10.3.

Variation: For low cholesterol mayonnaise, use 2 egg whites instead of 1 whole egg. Proceed as above.
NOTE: Due to the risk of raw eggs containing *Salmonella*, mayonnaise should not be tasted.

D. COOKED SALAD DRESSING (PERMANENT EMULSION)

2 tablespoon granulated sugar
1 teaspoon salt
2 tablespoons all-purpose flour
1 teaspoon dry mustard
Few grains of cayenne pepper

2 egg yolks, slightly beaten
3/4 cup milk
1/4 cup mild vinegar
1 tablespoon butter or margarine

1. Mix together in a small saucepan the sugar, salt, flour, dry mustard, and cayenne pepper.
2. Whisk in milk until the dry ingredients are dissolved. Whisk in egg yolks until blended.
3. Cook over moderate heat until mixture comes to a boil.
4. Remove from heat. Stir in butter and vinegar.
5. Record observations in Table 10.3.

Table 10.3 TABLE FOR THE EVALUATION OF EMULSIONS				
Product	Type of Emulsion	Stability	Appearance	Flavor

QUESTIONS

1. What is a permanent emulsion? Give an example.

2. What is a temporary emulsion? Give an example.

3. What is meant by an oil-in-water emulsion? Give an example.

4. What is meant by a water-in-oil emulsion? Give an example.

5. When preparing mayonnaise, what is meant by the dispersed phase?

6.	What is the difference between the emulsifier in mayonnaise and the emulsifier in the cooked salad dressing?

7.	What is the difference in fat content between mayonnaise and cooked salad dressing? What would be the nutritional quality of each?

8.	What is the function of dry mustard, paprika, and cayenne pepper in both the French Dressing and mayonnaise?

9.	a.	What causes the thickening of mayonnaise?

	b.	What may cause the emulsion in mayonnaise to break during preparation?

	c.	How can the emulsion be reformed during preparation?

10.	Oil and vinegar dressings require an oil that remains liquid when refrigerated. What kind of oil should be used?

LABORATORY 11

Gelatin

Gelatin

Gelatin is used primarily in food preparation to thicken and form a gel. There are two types of gelatin: coarse grind ("Knox" unflavored) and fine grind ("Jello"). These two types are dissolved differently, but the final results are achieved in a gel (liquid trapped in a solid). The purpose of this laboratory exercise is to introduce the student to the proper handling techniques for gelatin in food preparation and how it can be utilized and manipulated in a variety of products.

VOCABULARY

bromelin	foam	sol
Bavarian	gel	Spanish cream
chiffon	gelatin	sponge
collagen		

OBJECTIVES

1. To compare products made with unflavored gelatin and the available commercial mix.
2. To understand how to dissolve, melt, and disperse gelatin for product making.
3. To become familiar with the setting mechanism of gelatin.
4. To identify uses of gelatin in foams, molded salads, and desserts.

PRINCIPLES

1. Gelatin is derived from bones, skins, hides, and connective tissues (collagen) of animals.
2. There are two marketed forms of gelatin for home preparation: unflavored gelatin granules and flavored, sweetened granular gelatin.
3. The unflavored gelatin is coarse in granulation, while the flavored gelatin is fine or pulverized in granulation.
4. Since the gelatins are different in granulation, they are handled differently in the hydration process.
5. Unflavored or coarse gelatin is hydrated (also known as "softened" or "bloomed") in cold water initially, while the fine gelatin or flavored is placed in boiling water initially.
6. When gelatin is dissolved in water, the water molecules bond to the protein and the protein molecules swell slightly. Heating of the gelatin and the water forms a dispersion known as a sol (solid in a liquid).
7. Upon cooling, the hydrogen bonds come together, trap the water, and a gel is formed (liquid in a solid).
8. Slow cooling of the sol produces a stronger gel than rapid cooling, permitting a more ordered formation of the bonds between the protein chains. Setting of the gelatin begins at a temperature below 95°F.
9. Too much acid in the recipe can weaken the gel.
10. Too much sugar will also weaken the gel.
11. Proteolytic enzymes will prevent bond formation in the gelatin. Fresh pineapple contains bromelin and cannot be added to the gelatin mixture. Other food products that prevent gel formation are fresh kiwi and figs.
12. When adding solid ingredients, such as fruits, vegetables, and nuts, to the gelatin mixture, the gelatin mixture should be partially set (it has the appearance of unbeaten egg whites) to prevent flotation.
13. When preparing gelatin foams, the sol must be partially set (like unbeaten egg whites) before beating.
14. When the mold is completely gelled, loosen the sides of the gel with a knife, dip the mold into warm water for a few seconds to loosen, and then invert it onto a plate.
15. Gelatin is lacking in essential amino acid tryptophan.

I. TO COMPARE PRODUCTS MADE WITH UNFLAVORED GELATIN AND THE AVAILABLE COMMERCIAL MIX

1. Follow instructions on a 3 oz. package of a commercial gelatin dessert (orange flavor). Pour 1/2 of the dissolved mixture into a mold and chill. For the remainder, follow instructions for Orange Whip (Steps 2–5).
2. Prepare homemade orange gelatin.

3. Record the time for each set.
4. Rank the gels for stiffness, flavor, and clarity.

A. ORANGE GELATIN

1½ teaspoons unflavored gelatin   2 tablespoons lemon juice
2 tablespoons cold water   6 tablespoons orange juice
1/3 cup boiling water   1/4 cup granulated sugar

1. Hydrate (soften) the gelatin in cold water in a medium-sized bowl.
2. Add boiling water and stir to disperse and dissolve gelatin.
3. Add the lemon juice, orange juice, and sugar. Stir to dissolve sugar.
4. Pour into desired container or mold and place in refrigerator to set.
5. Record observations in Table 11.1.

B. ORANGE WHIP

1. Use the same ingredients as described in Section I.**A** and combine.
2. Place in the refrigerator to chill.
3. When the mixture is the consistency of thick egg white, whip with a hand mixer until the foam holds its shape.
4. Pour into a container or mold and return the mixture to the refrigerator to set.
5. Observe the amount of volume which increased due to beating.
6. Record observations in Table 11.1.

Table 11.1 TABLE FOR THE EVALUATION OF ORANGE GELATIN				
Type	Time to Set	Stiffness	Flavor	Clarity
Commercial				
Commercial whip				
Orange gelatin				
Orange whip				

QUESTIONS

1. Approximately how much gelatin is needed to gel 2 cups of liquid?

2. What effect does acid have upon a gelatin gel?

3. a. List the ways to speed the gelling of a gelatin mixture.

   b. Is this advantageous to the mixture?

4. Describe briefly the methods recommended for dispersing the two market forms of gelatin.

5. Janice knows that she cannot use fresh pineapple in a gelatin mold, but will she be able to use frozen pineapple to get the fresh fruit taste that she is looking for?

II. TO IDENTIFY AND TO BECOME FAMILIAR WITH THE DIFFERENT USES OF GELATIN

A. CARROT–PINEAPPLE SALAD

6 oz. can frozen orange juice concentrate, thawed
2/3 cup water
1 package (3 oz.) orange-flavored gelatin
2/3 cup lemon-lime-flavored carbonated beverage, chilled

1/8 teaspoon salt
1 can (8 1/4 oz.) crushed pineapple in syrup
1 cup carrot (1 medium), shredded

1. In a large saucepan, heat orange juice concentrate and water to boiling; stir in gelatin until dissolved.
2. Slowly add carbonated beverage; add salt.
3. Refrigerate until thickened but not set.
4. Fold in crushed pineapple and shredded carrots.
5. Pour into 4 cup mold.
6. Refrigerate until firm.
7. Unmold onto a lettuce lined serving plate.

B. SPANISH CREAM

1 tablespoon unflavored gelatin
1/4 cup cold milk
1 1/3 cups scalded milk
2 large eggs, separated

1/3 cup sugar
1/8 teaspoon salt
1 teaspoon vanilla extract

1. Place 1/4 cup milk in a custard cup; add the gelatin; let it stand while making the stirred custard.
2. In the top of a double boiler add the 1 1/3 cups of milk; heat the milk until it is scalded. Add 2 tablespoons of the sugar and salt. Mix until the sugar is dissolved.
3. In a small bowl whisk the 2 egg yolks; add 1/4 cup of the milk/sugar mixture to the beaten yolks and blend well. Then add this mixture back to the milk in the double boiler. Cook the mixture until the custard coats a spoon.
4. Add the hydrated gelatin to the hot custard. Stir until the gelatin is dissolved. Add the vanilla; chill until the mixture resembles unbeaten egg whites.
5. Beat the egg whites until they form soft peaks. Gradually add the remaining sugar with beating. Peaks should remain slightly soft.
6. Add the gelatin mixture and fold until just blended. Place in gelatin mold. Chill until firm.
7. Serve with whipped cream and/or raspberries.

C. PINEAPPLE–COCONUT BAVARIAN CREAM

2 teaspoons unflavored gelatin
1/4 cup cold water
1/4 cup sugar
1/2 cup sweetened shredded coconut, divided

1 cup crushed pineapple with syrup
1 tablespoon lemon juice
3/4 cup heavy cream, whipped

1. Put 1/4 cup cold water in a 2-quart mixing bowl. Add the gelatin; let it stand.
2. Drain syrup from pineapple into a 1 cup measure; add enough water to make 1/2 cup liquid. Bring mixture to a boil in a small saucepan or pot.
3. Add the boiling liquid to the gelatin; stir until the gelatin is dissolved.
4. Add the sugar and stir until dissolved; add lemon juice.
5. Chill the gelatin mixture until it is the consistency of raw egg white. Beat with an electric mixer until light and foamy.
6. Fold the whipped cream into the foamy gelatin mixture. Then fold in 1/4 cup of the coconut and the drained crushed pineapple. Do not over-fold. Pile the mixture into the mold; chill.
7. Toast the remaining shredded coconut by placing it in a Teflon or heavy skillet. Over moderate heat stir the coconut with a wooden spoon until the coconut is brown and toasted. Remove from skillet.
8. Unmold the Bavarian and sprinkle with the toasted coconut.

D. STRAWBERRY CHIFFON

1 tablespoon unflavored gelatin
1/4 cup cold water
1/4 cup sugar*
1 tablespoon lemon juice

1 pound frozen sliced or crushed strawberries, thawed
1 egg white
3/4 cup heavy cream, whipped

1. Put water in the upper part of a double boiler. Add gelatin; let it stand for 5 minutes.
2. Dissolve gelatin over boiling water. Add sugar, lemon juice, and berries; cook until all ingredients are heated.
3. Remove from heat. Chill mixture until it mounds from a spoon. Beat mixture with an electric mixer until it is light and foamy (soft peak stage). **Wash beaters before going to Step 4.**
4. Beat egg white to soft peak stage. Fold into gelatin mixture.
5. Fold whipped cream into the gelatin/egg white mixture. Pile into a mold; chill until firm.

*Increase sugar to 1/2 cup if fresh or unsweetened berries are used.

E. LIME BAVARIAN

2/3 cup evaporated milk (not the low fat or skim variety)
$1\frac{1}{2}$ oz. (42 g) lime gelatin (1/2 of a 3 oz. package)
2 tablespoons water
1 teaspoon unflavored gelatin
3/4 cup boiling water

1/2 cup sugar
2 tablespoons lime juice
1 tablespoon lemon juice
1 teaspoon grated lime rind
1 prepared chocolate wafer crumb crust (9 inch)
1 oz. semisweet chocolate, grated

1. Chill the evaporated milk in a 9-inch glass pie pan in the freezer until ice crystals are formed on the insides of the pan (30 minutes). **This is an important part of the recipe. If the milk is not cold it will not whip, therefore, the colder the better.**
2. In a custard cup add 2 tablespoons water and 1 teaspoon of the unflavored gelatin. Allow to soften for 5 minutes.
3. In a 2-quart bowl, blend lime gelatin and sugar; add boiling water and the softened gelatin; stir until gelatins and sugar are dissolved; add lime and lemon juices.
4. Chill the gelatin mixture until the mixture is the consistency of raw egg white.
5. Using an electric hand mixer beat the gelatin mixture to the soft peak stage. Add the grated lime rind during beating. (**Wash beaters before going to Step 6.**)
6. Place the chilled evaporated milk in a narrow 2-quart bowl. Beat at high speed with an electric mixer until it doubles in volume.
7. Add the beaten evaporated milk to the gelatin foam and beat with the electric mixer until the two are blended together.
8. Pour or spoon the mixture into the chocolate pie crust.
9. Chill. Just before serving, sprinkle the top with the grated chocolate.

F. WHITE COCONUT CHIFFON

1 tablespoon unflavored gelatin	1/4 teaspoon almond extract
1/4 cup all-purpose flour	4 large egg whites
1/2 teaspoon salt	1/2 cup granulated sugar
1/2 cup granulated sugar	1/4 teaspoon cream of tartar
1¾ cups 2% milk	1/2 cup flaked sweetened coconut
1 teaspoon vanilla	

1. In a 2-quart saucepan, blend together unflavored gelatin, flour, 1/2 cup sugar, and salt.
2. Slowly add milk and blend, making sure that no lumps are present.
3. Cook over medium heat, stirring constantly; bring mixture to a boil and remove from heat.
4. Add vanilla and almond extracts. Chill the mixture until it mounds slightly when dropped from a spoon.
5. Beat egg whites until foamy; add cream of tartar. Beat until soft peaks form. Add remaining sugar (1/2 cup), 1 tablespoon at a time. Beat until sugar is dissolved and egg whites form a stiff peak.
6. Fold cooled gelatin mixture into the meringue; fold in coconut.
7. Pour mixture into a baked crust or individual dessert dishes. Refrigerate until set.

Variation: For a special treat divide coconut and dye it with various food colors (e.g., red and blue against the white background).

QUESTIONS

1. Would you advise serving a tart molded salad on a hot day? Why?

2. What is the difference between a Bavarian and a chiffon?

3. Give an example how gelatin is used in dishes for different parts of a meal?

4. Since gelatin is derived from connective tissue, how important is its nutritional quality to the daily diet?

LABORATORY 12

Egg Cookery

LABORATORY 12
Egg Cookery

Eggs serve important functional roles in food preparation: foaming, coagulation, gelling, emulsification, binding, leavening, flavor, color, and nutritive value. The purpose of this laboratory exercise is to introduce the student to the different preparation techniques that are utilized with eggs.

VOCABULARY

candling	curdling	stirred custard
chalaza	denature	syneresis
coagulate	ferrous sulfide	vitelline membrane

OBJECTIVES

1. To illustrate factors which affect the quality of a cooked egg.
2. To illustrate how heat affects the gelation of egg proteins.
3. To observe the use of egg white foams in food preparation.
4. To become familiar with the different uses of eggs in food preparation.

PRINCIPLES

1. The basic parts of the egg are the shell, the white, and the yolk.
2. The fragile membrane, the **vitelline membrane**, covers the yolk, separating it from the white. Attached to opposite sides of the yolk are threads of thick white called **chalazae** that help keep the yolk centered in the white inside the egg (Fig. 12.1).

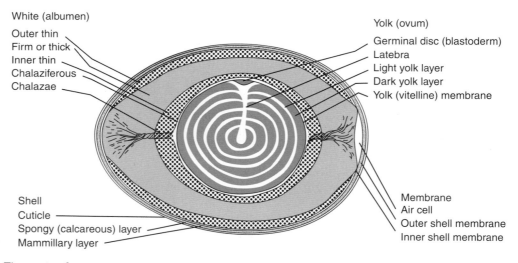

FIG. 12.1: The parts of an egg.

3. The shell of the egg is porous and allows gases and moisture to pass in and out.
4. Eggs are graded from highest to lowest quality: US Grade AA; US Grade A; and US Grade B.
5. Several changes occur in eggs over time:
   a. size of the air cell increases,
   b. carbon dioxide is lost; the egg becomes more alkaline,
   c. increase in alkalinity causes thinning of the white,
   d. yolk enlarges and becomes flatter due to the entrance of water from the white,
   e. flavor and aroma deterioration.

6. Eggs serve several functions in food preparation:
   a. Thickening agent in sauces and custards.
   b. Gelling agent in baked custard.
   c. Structural agent in baked products.
   d. Leavening agent; incorporating air with egg foams.
   e. Source of water in cookie dough and sponge cake.
   f. Emulsifier for mayonnaise, hollandaise sauce, cake batters, and cream puff dough.
   g. Binding agent in meat loaves; croquettes; for breading on meats.
   h. "Eggs" prepared in a variety of ways.
   i. Ingredients in many dishes where they serve no particular function other than to provide nutrition, flavor, texture, or color.
7. Heat affects the protein of an egg.
8. Coagulation and gelation are terms that are used to describe the effect of heat on eggs. **Gelation** is the formation of a gel structure (liquid trapped in a solid) by heat, as in baked custards. **Coagulation** is the change from the liquid to solid state with no particular structure (it occurs in a stirred custard and in fried and boiled eggs).
9. Eggs are a good source of complete protein, but also a good source of vitamin A, iron, riboflavin, and other vitamins and minerals. Eggs are also a rich source of cholesterol.
10. When whipped, raw egg white traps air pulled in by the beaters into bubbles surrounded by a film of protein molecules. Egg whites incorporate more air when beaten at room temperature.

I. TO OBSERVE THE EFFECT OF AGING ON THE RAW EGG

1. Use an egg candler to observe the shell and inside of an old egg and a fresh egg.
2. Break the eggs and observe on a flat plate, 6–7 inches in diameter.
3. Record your observations in Table 12.1.

Table 12.1 EVALUATION OF RAW EGGS		
Characteristic	Old	Fresh
Size of air sac		
Amount of thick white		
Amount of thin white		
Height of yolk		

QUESTIONS

1. Why was there an increase in the thinning of the white?

2. Why was the yolk flatter and not centered but at the edge of the white?

3. What type of preparation would be suitable for this type of egg?

II. TO BECOME FAMILIAR WITH VARIOUS METHODS TO COOK EGGS

A. POACHED EGGS

1. Fill pan with enough water to cover the egg.
2. Heat to simmering.
3. Carefully break a fresh egg and an old egg into separate custard cups or small flat plates. Slip one egg into the hot water.
4. Cover pan and keep water hot but below simmering. Cook until the white is firm and the yolk is of desired consistency (3–5 minutes).
5. Remove egg with a perforated spoon (use a slice of bread to rest the spoon on to soak up the excess water from the egg).
6. Repeat process with other egg.
7. Record observations in Table 12.2.

Variation: Affect of Vortex Action and Acid on Poached Egg

1. Break an egg into a custard cup.
2. Bring water to simmering. Stir water rapidly and quickly slip the egg into the vortex.
3. Add lemon juice or white vinegar (2 teaspoons to 1 quart of water). This tends to hasten the coagulation of the egg white.
4. Record observations in Table 12.2.

Table 12.2 EVALUATION OF POACHED EGGS		
Egg	Appearance of White	Appearance of Yolk
Old egg		
Fresh egg		
Fresh egg: vortex + lemon juice		

QUESTIONS

1. What effect did the acid have on the poached egg's appearance?

2. Compare the old egg versus the fresh egg as to appearance.

3. How long should the egg be poached in order to insure a microbiologically safe egg?

B. COOKED IN THE SHELL (HARD COOKED)

1. Place 2 eggs in a pan and cover with cold water.
2. Heat water to boiling.

3. Boil for:
   a. 10 minutes and remove 1 egg and cool rapidly under cold running water.
   b. Continue cooking the other egg for another 10 minutes (total cooking time: 20 minutes); remove from boiling water, but allow to cool to room temperature.
   c. Crack and peel the shell off the cooked eggs. Cut in half to observe the color and texture of the yolk.

Variation

Bring 2 eggs, 1 teaspoon salt, and 1 quart of water to a boil in a medium saucepan over high heat. As soon as the water reaches a boil, remove the pan from the heat, cover, and let sit for **exactly 10 minutes**. Meanwhile, prepare a bowl of ice water. Transfer the eggs to the ice water and let cool for 5 minutes. Peel the eggs and split in half to observe the color and texture of the yolk. Record observations in Table 12.3.

Table 12.3 EVALUATION OF HARD BOILED EGGS				
Treatment	Presence of Sulfur Ring	Texture of Yolk	Texture of White	Odor
10 minutes and cooled				
20 minutes and no cooling				
Variation: no boiling				

QUESTIONS

1. Which method produced the most attractive hard-cooked egg and why?

2. What caused the green ring around the yolk of the egg? What is the name of the ring and how can it be avoided?

C. FRIED EGG

Variation 1: Fat Only

1. Heat 1 tablespoon of butter or margarine in an 8-inch skillet over moderate heat.
2. Break egg into a flat dish, and then slide the egg into the heated pan.
3. Reduce heat slightly and cook, basting the egg with the melted fat until desired firmness and the yolk will develop a film over itself.
4. Record observations in Table 12.4.

Variation 2: Fat and Water

1. Heat 2 teaspoons of butter or margarine in an 8-inch skillet over moderate heat.
2. Break an egg onto a small flat plate and then slide the egg into the heated fat.
3. Add 2 tablespoons of water to the pan and cover (the steam helps to form a coating over the yolk).
4. Reduce heat and cook on a low temperature until the white has coagulated.
5. Record observations in Table 12.4.

Table 12.4 EVALUATION OF FRIED EGGS				
Variation	Appearance of Yolk	Texture of Yolk	Texture of White	Flavor
1				
2				

D. SCRAMBLED EGGS

2 large eggs   few grains salt
2 tablespoons milk or cream   1 teaspoon fat

1. Beat eggs with milk and salt with a whisk. There should be streaks of yolk and white.
2. Melt fat in an 8-inch skillet over moderate heat.
3. Add egg mixture and with the back of a spatula stir the egg lifting the watery part until the egg has coagulated. Stir gently until the entire mass is coagulated but soft.
4. Record observations in Table 12.5.

E. SCRAMBLED EGGS—MICROWAVE OVEN METHOD

2 large eggs   1/8 teaspoon salt
2 tablespoons milk   1 teaspoon butter or margarine

1. In a small bowl, blend together the eggs, milk, and salt until the yolks and whites are broken up, but not foamy.
2. Place butter into a 2-cup microwave safe cooking container. Microwave on HIGH for 15 seconds or until the butter is melted.
3. Pour the beaten eggs into the container and return the container to the microwave. Microwave on HIGH for 30 seconds; stir. Microwave again on HIGH for another 30 seconds; stir.
4. Microwave for another 30 seconds at HIGH; at this point the eggs will appear moist and slightly creamy; they are servable as the eggs will continue to cook for a short time ("standing time"). However, the eggs can be cooked to a greater degree of doneness by heating at 5-second intervals. Care must be taken to avoid overcooking.
5. Record observations in Table 12.5.

F. SCRAMBLED "EGGBEATERS" (COMMERCIAL EGG SUBSTITUTE)

1/2 cup eggbeaters   1 teaspoon margarine

1. Melt margarine in an 8-inch skillet over medium heat.
   When margarine is melted and hot pour in the eggbeaters; **do not stir**.
2. When the eggbeaters begin to get firm around the edges, push set portion to the center of the pan, allowing the uncooked portion to flow to the edges of the pan. **Do not stir**. Turn over to cook to desired doneness. Season eggbeaters as desired.
3. Record observations in Table 12.5.

Table 12.5 EVALUATION OF SCRAMBLED EGGS			
Type	Moistness	Texture	Color
Conventional			
Microwaved			
Eggbeaters			

CHARACTERISTICS OF HIGH-QUALITY SCRAMBLED EGGS

Appearance: Even masses appear slightly moist and creamy; usually egg masses are large.
Consistency: Even consistency throughout; all liquid is held by a coagulated protein.
Tenderness: Egg masses are tender and have little resistance when cut or chewed.
Flavor: Mild egg flavor; fat used will enhance (e.g., butter, bacon).

G. FRENCH OMELET (PLAIN OMELET)

1. Use recipe for scrambled eggs.
2. Lift cooked eggs with spatula and let uncooked liquid run to the bottom of the skillet. Keep omelet uniformly thick.
3. Fold the omelet in half in the skillet and turn onto a plate to serve.
4. Record observations in Table 12.6.

H. PUFFY OMELET

4 large eggs, separated   1/4 teaspoon salt
1/4 cup water   1 tablespoon butter or margarine
1/4 teaspoon cream of tartar

1. Preheat oven to 350°F.
2. Beat egg yolks until thick and lemon colored, about 5 minutes.
3. Add water, cream of tartar, and salt to whites in a medium-sized mixing bowl. Beat whites until stiff but not dry or until whites no longer slip when bowl is tilted.
4. Fold beaten yolks into whites.
5. On medium high heat, melt butter in a 10-inch skillet or omelet pan with an ovenproof handle until just hot enough to sizzle a drop of water.
6. Pour in omelet mixture; level surface gently. Reduce heat to medium. Cook slowly until puffy and lightly browned on bottom, about 5 minutes.
7. Lift omelet at edge to judge color. Place omelet in preheated oven for 10–12 minutes or until a knife inserted halfway between the center and outside edge comes out clean.
8. Record observations in Table 12.6.

Table 12.6 EVALUATION OF OMELETS			
Omelet	Texture	Appearance	Flavor
Plain omelet (French omelet)			
Puffy omelet			

QUESTIONS

1. Discuss the effect of cooking method on the flavor and caloric content of fried eggs.

2. Distinguish between scrambled eggs, French omelet, and puffy omelet.

3. a. What is the difference in the appearance and flavor of the scrambled eggs and the eggbeaters?

   b. Compare the nutritional differences between these two egg products.

4. What precautions should be taken when using the microwave when cooking eggs? What effect does heat have on the quality of the cooked egg?

III. TO UNDERSTAND HOW HEAT AFFECTS GELATION OF EGG PROTEINS

A. BAKED CUSTARD

3 eggs, slightly beaten
1/3 cup sugar
1 teaspoon vanilla

Dash of salt
2$\frac{1}{2}$ cups milk, scalded
Ground nutmeg

1. Heat oven to 350°F.
2. Mix eggs, sugar, vanilla, and salt in a medium-sized bowl. Gradually stir in milk
3. Pour into six 6 oz. custard cups. Sprinkle with nutmeg.
4. Place cups in a rectangular pan, 13 × 9 × 2 inches on oven rack. Pour very hot water into pan to within 1/2 inch of tops of cups.
5. Bake custard about 45 minutes or until a knife inserted halfway between center and edge comes out clean.
6. Remove custard cups from the hot water or they will continue to cook. Serve the custard warm or cold. Refrigerate any remaining custard.
7. Record observations in Table 12.7.

B. STIRRED CUSTARD WITH FLOATING ISLAND

1$\frac{1}{2}$ cups milk
4 large egg yolks
1/4 cup granulated sugar

1/4 teaspoon salt
1 teaspoon vanilla

1. Heat milk in top of a double boiler over boiling water.
2. Beat egg yolks in a small bowl. Blend in sugar and salt.
3. Gradually stir in hot milk. Return to double boiler. Cook the custard over **simmering water, stirring constantly with a metal spoon**.
4. When the custard coats the spoon (**thin coating**), remove from heat. **Another method to determine doneness: turn the spoon over and run your finger down the back of the spoon. A clean path should be left.**
5. Immediately remove the custard from the double boiler to cool quickly.
6. Blend in vanilla.
7. Record observations in Table 12.7.

Meringue (Floating Island)

2 large egg whites
1/4 cup granulated sugar

1/4 teaspoon vanilla extract

1. Preheat oven to 350°F.
2. Place egg whites in a small mixing bowl. Add vanilla and beat the egg whites to the soft peak stage.
3. Gradually add the sugar, and beat the meringue until stiff peaks form and there is no trace of sugar when the meringue is tested by rubbing a small amount between your fingertips.

4. Pour 1/2 inch hot water into a 13 × 9 × 2 inch pan.
5. Drop the meringue from a teaspoon onto the hot water to form approximately 10 "islands."
6. Place the pan in the oven and bake the meringue islands until the tips are golden brown, approximately for 20 minutes.
7. Use a slotted spoon or pancake turner to remove the islands from the water and onto a plate to cool.
8. Pour stirred custard (from above) onto a plate with a high edge (a pie plate works well), and place the cooled islands on the custard; serve. Refrigerate any leftovers.

C. LOW-FAT STIRRED CUSTARD

2 cups skim evaporated milk
3 tablespoons light brown sugar, packed

2 large eggs, thoroughly beaten
1 teaspoon vanilla extract

1. Heat the water in the bottom of a double boiler.
2. In the top part add the milk and the brown sugar.
3. Place over the hot water, and scald the milk.
4. Lower the temperature of the water to simmer.
5. Add a small amount of the scalded milk to the beaten eggs; stir and mix thoroughly and add this back to the milk in the double boiler. Mix and incorporate thoroughly.
6. Place the double boiler back over the simmering water and continue to cook, stirring constantly until mixture coats a spoon.
7. Immediately remove from heat and pour into a bowl. Add vanilla.
8. Record observations in Table 12.7.

Suggestion. Excellent over fresh fruit: sliced strawberries, raspberries, and sliced peaches.

Table 12.7 EVALUATION OF CUSTARDS				
Type	Firmness of Gel	Syneresis	Flavor	Color
Baked				
Stirred				
Stirred/floating island				
Stirred low-fat				

QUESTIONS

1. What are the main ingredients in custard?

2. What is the difference between a baked custard and a stirred custard?

3. What are the main quality characteristics of a well-made baked custard and a stirred custard?

4. What is the difference between determining the cooking end point of a baked custard and a stirred custard?

5. What happens to a baked custard and a stirred custard when they are overcooked?

6. What is the difference in the setting mechanism of the egg in the baked custard and the stirred custard?

7. Why was water used in cooking the baked and stirred custards?

8. Why must both time and temperature be controlled in cooking custard mixtures?

IV. TO OBSERVE THE USE OF EGG WHITE FOAMS IN FOOD PREPARATION

A. CHEESE SOUFFLE

2 large egg yolks
3 large egg whites
2 tablespoons all-purpose four
1/4 teaspoon salt

Dash of cayenne pepper
2/3 cup milk
1 tablespoon butter or margarine
3/4 cup mild or sharp Cheddar cheese, shredded

1. Preheat the oven to 350°F. Separate eggs. Lightly grease the bottom of a 1-quart soufflé dish.
2. Blend flour, salt, and cayenne pepper together in a 1-quart saucepan. Whisk in the milk until all the ingredients are dissolved.
3. Place saucepan on heat and bring the mixture to a boil with constant stirring. Boil the sauce for 1 minute and then remove it from the heat.
4. Add the butter and the shredded cheese to the hot sauce. Stir until the cheese has melted.
5. Stir to combine the yolks; add a small amount of the cheese sauce to the yolks; blend; add the yolk mixture to the cheese sauce and mix well.
6. Beat whites until the peaks are stiff but still shiny.
7. Add a small amount (approximately 1/4 cup) of the beaten whites to the cheese sauce and stir in. This will help to lighten the sauce and will help in folding in the whites.
8. Add the cheese sauce to the beaten whites and fold into the whites.
9. Pour the mixture into the soufflé dish. Set the dish in a pan of warm water—**water should be the same depth as the amount of soufflé in its baking dish**.
10. Place the soufflé in the preheated oven and bake until a knife comes out clean when inserted in the center, about 50 minutes. Serve immediately.

B. CHOCOLATE SOUFFLE ROLL

1 package (6 oz.) semisweet chocolate morsels
2 tablespoons prepared black coffee
6 large eggs, separated
1/2 cup granulated sugar, divided

2 teaspoons vanilla extract, divided
1 tablespoon cocoa
5 tablespoons confectioners' sugar
1 cup heavy cream

1. Preheat the oven to 350°F. Grease bottom and sides of a jelly roll pan, 15 × 10 × 1 inch, with vegetable spray. Line with waxed paper just on the bottom of the pan and let the waxed paper over hang at both opposite ends of the pan. Spray the waxed paper with the vegetable spray; set aside.
2. Place the chocolate morsels and prepared coffee in top of a double boiler; bring the water to a boil. Reduce heat to low; stir occasionally, until chocolate melts. The microwave could also be used: place chocolate morsels and coffee in a microwave safe bowl and microwave on HIGH for 30 seconds. Take bowl out and stir. Return bowl back to the microwave and microwave on HIGH for 15 seconds; stir; repeat procedure until chocolate is melted.

3. Beat egg yolks in a large mixing bowl at high speed with an electric mixer until foamy. Gradually add **1/4 cup of the granulated sugar**, beating until mixture is thick. Stir in 1 teaspoon vanilla extract.
4. Gradually stir in melted chocolate. Make sure chocolate is well distributed and there are no yellow streaks.
5. Beat egg whites until soft peaks form. **Gradually add the remaining 1/4 cup granulated sugar**.
6. Stir a small amount of the egg white meringue into the chocolate mixture. This "lightens" the chocolate mixture in that it is no longer heavy and allows for a better incorporation of the meringue without deflating it.
7. Fold the remainder of the meringue into the chocolate mixture.
8. Pour the chocolate mixture into the prepared pan, spreading evenly. Bake the soufflé for 15–18 minutes on the middle position in the oven. Do not overbake.
9. Remove the pan from the oven and immediately place a damp towel on top of the soufflé to cover. Allow the soufflé to cool at least 1 hour. Soufflé will sink. That is fine.
10. Carefully remove towel and loosen edges with a metal spatula.
11. Before removing the soufflé from the pan, place two lengths of waxed paper (longer than the jelly roll pan) on a smooth slightly damp surface. Tape the waxed paper together (overlap the two pieces, then tape). Turn the joined waxed paper over and tape the other side. Sift over the waxed paper with 1 tablespoon of cocoa and 3 tablespoon confectioners' sugar. Add more if you wish. This prevents the soufflé from sticking when rolling.
12. Beat heavy cream with **2 tablespoons confectioners' sugar** and **1 teaspoon vanilla extract**; beat until stiff peaks are formed.
13. Quickly invert jelly roll pan onto the prepared waxed paper with long side nearest you. Remove pan and carefully peel waxed paper from the chocolate roll.
14. Spoon whipped cream over chocolate roll; spreading so that there is more on the side facing you (mixture will spread out as you roll); leave a 1-inch margin on all sides.
15. Starting at the long side, carefully roll jelly roll fashion using the waxed paper to help support the soufflé as you roll. Secure waxed paper around the soufflé; smooth and shape it with your hands.
16. On your last roll have a flat sheet ready to place the rolled soufflé seamed side down onto the sheet. Refrigerate until ready to use.
17. Before serving, sift with additional confectioners' sugar.

Variation

In place of the vanilla whipped cream, substitute chocolate whipped cream.

Chocolate Whipped Cream

1 cup heavy cream	1/4 cup sifted natural unsweetened cocoa
1/2 cup confectioners' sugar	1/2 teaspoon vanilla extract

1. Combine all ingredients in a medium-sized bowl. Refrigerate and cover for 30 minutes.
2. With portable electric mixer at high speed, beat mixture until stiff. Refrigerate until ready to use.

QUESTIONS

1. Why are soufflés and baked custards cooked in a pan of water?

2. Explain the effect of overbeating egg white foam to be used in a soufflé.

3. What type of white sauce is used in a soufflé?

4. What precautions are necessary when eggs are combined with a hot mixture?

5. Why should egg whites be used immediately after they are beaten?

6. What is the difference between a soufflé and a puffy omelet?

V. MISCELLANEOUS EGG COOKERY

A. MANICOTTI

Crepe Batter

6 large eggs
3/4 cup water
3/4 cup milk

$1\frac{1}{2}$ cups all-purpose flour
1/4 teaspoon salt

Cheese Filling

2 pounds skim ricotta cheese
1/2 cup grated Parmesan cheese
2 large eggs

2 tablespoons minced parsley
Dash salt and pepper

1. *Prepare crepe batter:*
    a. In a blender jar add the eggs, water, milk, flour, and salt.
    b. Blend all ingredients until smooth.
    c. Allow batter to "rest" in the refrigerator for 30 minutes.
2. *Prepare cheese filling:*
    a. Mix all ingredients in a medium-sized bowl.
    b. Refrigerate until ready to use.
3. *Preparing crepes:*
    a. Take an 8-inch nonstick omelet pan and spray it with vegetable spray.
    b. Measure out 3 tablespoons of batter into custard cups and set aside. Also, cut pieces of waxed paper that will hold the individual crepe when cooked.
    c. Heat pan over medium heat. Test hotness of pan by sprinkling with water: water should "dance" on surface of pan.
    d. Quickly pour in batter and rotate the pan to evenly distribute the batter. Cook crepe on medium to low heat.
    e. Crepe is ready when it feels dry and pulls away from the sides of the pan.
    f. Place each crepe on a piece of waxed paper. After the crepe has cooled for about 5 minutes lift the crepe from the waxed paper to make sure it is not stuck and will make for easy rolling when filled.
    g. Continue making crepes until batter is finished (approximately 16–18 crepes).
4. *Assembling crepes:*
    a. Preheat oven to 375°F. Spread 1/4 cup cheese filling down the center of each crepe. Roll up the crepe and place it seam side down in a lightly greased 13 × 9 × 2 inch pan. The pan should have some tomato sauce on the bottom. **Use a prepared tomato sauce**.
    b. Continue layering the crepes (manicotti) in a single layer. **You will need two pans**. Pour tomato sauce over the top of the manicotti. Sprinkle with Parmesan cheese.
    c. Cover pans with aluminum foil and bake for 30 minutes.
    d. Remove foil and bake for another 5–10 minutes until bubbly.

QUESTIONS

1. What is the difference between a crepe and a pancake?

2. What is the functional role of the egg in the crepe batter?

3. a. What is meant by a seasoned pan?

   b. How does this pan affect egg cookery?

LABORATORY 13

Milk and Cheese

Milk is homogenized and pasteurized. There are a variety of milk products available to the consumer. However, the main proteins in these milk products are casein and whey. These proteins affect the cooking quality of milk. Cheese is made from milk. This laboratory exercise will introduce the student to the different types of milk and cheese products as well as their role in food preparation.

VOCABULARY

casein	evaporated milk	ripened cheese
Cheddar	homogenization	soft cheese
cheese food	lactoglobulin	sweet acidophilus milk
cheese spread	lactalbumin	sweetened condensed milk
cultured milk product	pasteurization	unripened cheese
curd	processed cheese	whey
enzyme	rennin	

OBJECTIVES

1. To understand how different proteins found in milk are affected by heat and acid.
2. To become familiar with the different milk and cheese products available.
3. To observe how milk and cheese behave during cooking.

PRINCIPLES

1. Casein makes up about 80% of the protein in milk, while whey makes up 20%.
2. Whey proteins are made up of lactalbumin and lactoglobulin.
3. Homogenization breaks up the fat globules in the milk into very small particles that remain dispersed evenly throughout the milk. Homogenization forms a permanent emulsion (oil-in-water emulsion).
4. When milk is heated on a surface unit or range, the whey proteins precipitate and settle on the bottom of the pan.
5. Casein is precipitated by acid. Precipitation of the casein in milk is desirable in making cheese and cultured milk products.
6. Rennin is an enzyme which also precipitates casein in milk.
7. Specific conditions must be met for the rennin precipitation of casein:
   a. active at 104–108°F (40–42°C).
   b. inactivated above 140°F (60°C).
   c. optimal pH 5.8–6.4.
   d. free calcium ions from milk are required by the enzyme to precipitate the enzyme.
8. In the commercial production of cheese:
   a. a combination of acid and rennin precipitation is used.
   b. pasteurized milk, cream, non-fat, or combination are warmed and inoculated with the desired lactic acid producing bacteria.
   c. after sufficient acid has been produced to yield a pH of 5.8, milk is inoculated with rennin.
   d. the renin causes the milk to clot or gel.
   e. the gel is cut and drained.
   f. the gel is heated slightly to shrink the curd and expel the whey.
   g. the whey is then separated from the curds.
9. Unripened cheese is made of curds with no other treatment except for the addition of small amount of salt and sometimes cream.
10. Ripened cheese is made of curds that are inoculated with the desired bacteria and/or mold that gives the cheese its characteristic flavor, aroma, and texture.
11. Processed cheese is a blend of one or more natural cheeses that have been heated or pasteurized. Emulsifiers and water are added, and the mixture is whipped to form a smooth, homogeneous product. Cheese food and

cheese spread are made by the same method as processed cheese; however, they have less fat and more moisture, respectively.

12. When hard cheeses, such as Cheddar, Swiss, or Monterey Jack, are heated, they first soften as the fat melts; the higher the fat content, the more readily the cheese liquefies. Continued heating at too high a temperature or for too long a time causes the cheese to lose moisture, shrink, and toughen.

13. Processed cheese is more stable to heat than natural cheese. The emulsifiers in processed cheese improve its blending properties.

I. TO BECOME FAMILIAR WITH VARIOUS MILK PRODUCTS

A. EVALUATE VARIOUS MILK AND MILK PRODUCTS

Instructions: Various milk and milk products have been selected for evaluation. Please sample the various products in the cups provided and record your observations in Table 13.1.

Table 13.1 EVALUATION OF MILK AND MILK PRODUCTS			
Milk	Appearance	Flavor	Acceptability

II. TO SHOW THE EFFECTS OF TEMPERATURE UPON THE CLOTTING OF MILK BY RENNIN

Temperature	Amount of Milk
42°F	1/2 cup
105°F	1/2 cup
212°F	1/2 cup

1. Adjust temperature of 1/2 cup milk to each of the stated temperatures.
2. Dissolve 1/4 tablet of rennet in 1 tablespoon cold water.
3. Add rennet solution to milk. Stir quickly.
4. Allow it to stand at room temperature for 10 minutes.
5. Refrigerate for 10 minutes and record results in Table 13.2.

Table 13.2 OBSERVATIONS OF MILK CLOTTING BY RENNIN	
Temperature	Observation of Clot Formation
42°F	
105°F	
212°F	

QUESTIONS

1. What is the optimum temperature for the coagulation of milk by rennin?

2. What is the reaction that is involved?

3. List the requirements for the precipitation of casein by rennin.

III. TO STUDY THE EFFECTS OF HEAT AND ACID ON MILK PROTEINS

A. COAGULATION OF MILK BY HEAT

1. Place 1/2 cup whole milk in a small saucepan; place over low heat.
2. Heat slowly to boiling. Do not stir. Remove from heat.
3. Record observations in Table 13.3.

B. COAGULATION OF MILK BY ACID

1. Place 1 cup whole milk minus 1 tablespoon in a glass measuring cup.
2. Measure the pH:_____.
3. Add 2 tablespoons vinegar. Stir; let it stand for 10 minutes.
4. Measure the pH:_____.
5. Record observations in Table 13.3.

C. COAGULATION OF SWEETENED CONDENSED MILK BY HEAT

1. Preheat oven to 425°F. Pour 1 can (14 oz.) sweetened condensed milk into an 8- or 9-inch glass pie plate. Cover with aluminum foil.
2. Pour hot water about 1/4 inch in a large shallow baking pan. Place covered pie dish in the pan of water.
3. Place the pan of water with the covered pie dish in the preheated oven. Bake for 1 hour and 30 minutes or until condensed milk is thick and caramel color (add hot water to pan as needed during the baking process).
4. Remove pan from oven. Allow pan to cool at least 15 minutes before removing the pie plate from the water bath. When the pie plate is removed from the water bath, take off the aluminum foil.
5. Record observations in Table 13.3.

D. COAGULATION OF SWEETENED CONDENSED MILK BY ACID

1/2 cup sweetened condensed milk   1/4 cup freshly squeezed lemon juice

1. Gradually stir lemon juice into sweetened condensed milk.
2. Allow at least 10 minutes before cutting into it.
3. Record observations in Table 13.3.

   a. *Key Lime Pie*:

   1 nine-inch prepared graham cracker crust or vanilla wafer crust
   1 (14 oz.) sweetened condensed milk (or the low-fat variety)

   8 oz. of cream cheese or Neufchatel cheese (room temperature)
   1 teaspoon grated lime rind
   7 tablespoons key lime juice
   2–3 drops of green food color, optional

1. In a food processor* combine all the filling ingredients: sweetened condensed milk, cream cheese, grated lime rind, lime juice, and food color.
2. Beat until smooth.
3. Pour the filling into the prepared pie shell and refrigerate at least 3 hours (for faster setting, place the pie covered in the freezer for at least 1 hour) until set.
4. Refrigerate any leftovers.

*If food processor is not available, place ingredients in a medium-sized mixing bowl and beat ingredients with a portable electric mixer until all ingredients are combined and the mixture is smooth.

Table 13.3 EVALUATION OF HEAT AND ACID EFFECT ON MILK PROTEINS				
Milk	Heat	Acid	Appearance	Protein(s) Affected
Whole milk				
Whole milk				
Sweetened condensed milk				
Sweetened condensed milk				

QUESTIONS

1. a. What effect did heat have on whole milk?

   b. What protein was affected?

2. a. What effect did acid have on whole milk?

   b. Which protein is found in the watery part?

c. Which protein is found in the curd?

3. a. What effect did acid and heat have on the sweetened condensed milk?

b. Was it the same effect observed with the whole milk?

c. Explain why if there were differences.

d. What browning reaction occurs when the sweetened condensed milk is processed and during the heating in the oven?

IV. TO OBSERVE HOW MILK BEHAVES DURING PREPARATION

A. CREAM OF TOMATO SOUP

2/3 cup tomato juice
1 whole peppercorn
1/4 bay leaf
1 small onion, thinly sliced
1 tablespoon tomato paste

1 tablespoon butter or margarine
1 tablespoon all-purpose flour
1/4 teaspoon salt
1/2 cup whole milk, skim milk, or half and half

1. Simmer together for 5 minutes the tomato juice, peppercorn, bay leaf, and onion. Strain into a glass measuring cup.
2. Add tomato juice to bring the liquid back to 2/3 cup. Add tomato paste and blend.
3. Make a white sauce by melting the butter in a small saucepan, adding the flour and salt, and blend until smooth. Remove from heat.
4. Slowly add the milk to the tomato mixture. Add this mixture to the roux.
5. Return the saucepan to the heat; bring to a boil with constant stirring.
6. Record observations in Table 13.4.

Table 13.4 EVALUATION OF CREAM OF TOMATO SOUP				
Variation	Appearance	Consistency	Color	Flavor
Whole milk				
Skim milk				
Half and half				

137

QUESTIONS

1. Describe methods to reduce or eliminate curdling of milk mixtures that contain acidic ingredients.

2. What ingredient serves as a stabilizing agent in the soup recipe?

3. What step or steps are important in the recipe to prevent curdling of the soup?

4. What is the desirable consistency for a cream soup?

V. TO DIFFERENTIATE AMONG THE MANY AVAILABLE VARIETIES OF CHEESE

Sample varieties of cheese are provided. Classify them under the appropriate headings and record other information in Table 13.5.

Table 13.5 EVALUATION OF CHEESE VARIETIES						
Classification	Name	Flavor	Texture	Odor	Color	Uses
Unripened (low fat)						
Unripened (high fat)						
Bacteria ripened (soft)						
Bacteria ripened (semi-hard)						
Bacteria ripened (hard)						
Mold ripened (soft)						
Mold ripened (semi-hard)						
Processed cheese						
Cheese food or spread						

VI. TO COMPARE HOW VARIOUS PROCESSED CHEESE PRODUCTS MELT AS COMPARED TO A NATURAL CHEESE PRODUCT

1. Set out 4 slices of bread.
2. Place cheese to be tested on a separate bread slice.
3. Place bread slices on a broiler pan and place pan 5 inches away from under a preheated broiler.
4. Watch carefully and record time and characteristics of the melted cheese in Table 13.6.

Table 13.6	EFFECT OF HEAT ON MELTING PROPERTIES OF CHEESE					
Type of Cheese	Time to Melt	Smoothness	Consistency	Flavor	Observations
Natural					
Processed cheese					
Processed cheese food					
Processed cheese spread					

QUESTIONS

1. Regarding the results of the processed cheese, account for the differences, if any, in the

    a. melting time

    b. smoothness

    c. consistency

    d. flavor

2. What would account for flavor differences between unripened and ripened cheeses?

VII. TO UNDERSTAND HOW TO COOK WITH CHEESE

A. MACARONI AND CHEESE

2 cups uncooked elbow macaroni, cooked
   according to package directions
1/4 cup all-purpose flour
1/2 teaspoon salt
1/8 teaspoon white pepper

1/2 teaspoon dry mustard
$2^1/_2$ cups 2% milk
$2^1/_2$ cups shredded Cheddar cheese, mild or sharp
1 tablespoon Worcestershire sauce
1 tablespoon finely grated yellow onion

1. Preheat oven to 350°F.
2. In a 2-quart saucepan combine the flour, salt, white pepper, and dry mustard.
3. Whisk in milk until all ingredients are dissolved.
4. Over moderate heat, cook milk mixture, stirring constantly, until mixture boils.
5. Remove cream sauce from the heat, and stir in 2 cups of the Cheddar cheese. Stir mixture until the cheese melts. Stir in the Worcestershire sauce and the grated onion.
6. Add the cooked elbow macaroni and mix thoroughly until the cheese sauce coats the macaroni.
7. Place the coated macaroni into a greased 2-quart casserole dish. Sprinkle top of the macaroni and cheese with the reserved 1/2 cup shredded cheese.

8. Bake uncovered for about 30 minutes, until bubbly and lightly browned.
9. Record observations in Table 13.7.

Variation

Follow the above recipe, except substitute either:

1. Low-fat Cheddar cheese
   OR
2. Non-fat Cheddar cheese

B. MACARONI WITH FOUR CHEESES

2 cups uncooked elbow macaroni, cooked
   according to package directions
1/4 cup all-purpose flour
1/2 teaspoon salt
1/8 teaspoon white pepper
1/2 teaspoon dry mustard
2$\frac{1}{2}$ cups 2% milk
3/4 cup shredded Swiss cheese

1/2 cup shredded Parmesan cheese
1 cup shredded extra sharp Cheddar cheese
3 oz. shredded processed cheese product
   (Velveeta)
1 tablespoon Worcestershire sauce
1 tablespoon finely grated yellow onion
1/2 cup crushed onion flavored croutons

1. Preheat oven to 350°F.
2. In a 2-quart saucepan combine together the flour, salt, white pepper, and dry mustard.
3. Whisk milk into dry ingredients until they are dissolved.
4. Over moderate heat, cook milk mixture, stirring constantly, until the mixture thickens and boils.
5. Remove pan from the heat and stir in the Cheddar cheese, Swiss cheese, Parmesan cheese, and the Velveeta cheese until all are melted.
6. Stir in Worcestershire sauce and grated onion; add the cooked elbow macaroni and mix thoroughly until the cheese sauce coats the macaroni.
7. Spoon mixture into a greased 2-quart casserole. Sprinkle the crushed croutons on top of the macaroni and cheese.
8. Bake uncovered for about 30 minutes or until bubbly.
9. Record observations in Table 13.7.

Table 13.7 EVALUATION OF MACARONI AND CHEESE				
Variation	Appearance	Texture	Flavor	Comments
Natural cheddar				
Low-fat cheddar				
Non-fat cheddar				
Four cheese				

GENERAL QUESTIONS

1. How is cheese ripened or cured?

2. What determines the fat content of cheese?

3. How is processed cheese prepared?

4. When cooking with cheese, what precautions should be taken with regard to heating?

5. What nutritional contribution does cheese make to the diet?

6. How does milk pasteurization affect the quality of cheese, especially safety?

LABORATORY 14

Meat and Poultry

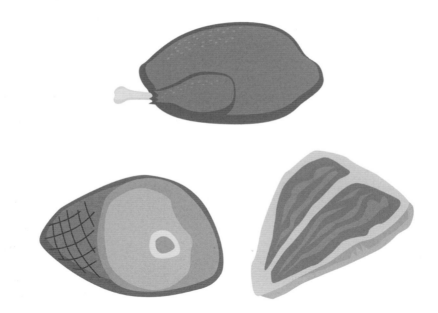

LABORATORY 14
Meat and Poultry

Cooking of a particular cut of meat requires the knowledge where the piece was derived from the carcass. This will influence whether moist or dry heat should be used. With poultry, however, depending on the age and size of the chicken, either moist or dry heat can be used due to the presence of connective tissue. This laboratory exercise will introduce the student to the selection of different meat cuts and poultry, as well as the varied preparation techniques.

VOCABULARY

actin	gelatin	oxymyoglobin
braising	marbling	stewing
collagen	metmyoglobin	trichinosis
connective tissue	muscle fiber	0157:H7
elastin	myoglobin	
fricassee	myosin	

OBJECTIVES

1. To learn the different characteristics of beef, pork, veal, and lamb.
2. To learn to identify the different cuts of meat and the type of cooking procedures associated with each.
3. To learn to differentiate between the various market forms of poultry and their proper preparation.
4. To learn safe handling practices and proper end-point temperatures when preparing meat and poultry products.

PRINCIPLES

1. The nature and proportions of muscle fiber, connective tissue, and fatty tissue will directly affect the eating and cooking quality of meats.
2. The type of bone will also be an indication of the tenderness and the type of cooking required for a piece of meat; for example, a T-bone steak (dry heat) as opposed to a blade steak (moist heat).
3. Muscle fiber is made up of thick filaments composed of myosin protein and thin filaments which make up the actin protein.
4. The connective tissue is made up of collagen which is the principal connective tissue in meat. It is flexible but lacks the elasticity of elastin. Collagen breaks down to gelatin when subjected to moist heat over a prolonged cooking period, while elastin is unaffected.
5. Myoglobin is the pigment in meat that is responsible for its color. When exposed to oxygen, the pigment turns a bright "cherry red" color which is known as oxymyoglobin. After a while the color starts to take on brown overtones and this pigment is known as metmyoglobin.
6. Liver, heart, tongue, brains, sweetbreads, and tripe from various animals are classified as organ or variety meats. Some require dry heat, such as, liver, brains, and sweetbreads; while others require moist heat, such as tongue and tripe.
7. Meat can be tenderized by
   a. grinding.
   b. use of proteolytic enzymes (bromelin, ficin, or papain).
   c. pounding.
   d. marinating.
   e. cutting against the grain.
8. The most important reason to cook meat is to make it tender by softening the collagen connective tissue. Excessive heat, however, can actually toughen meat because heat causes the muscle fibers in the lean portion of the muscle to shrink and lose water.
9. Overcooking makes meat tough, rubbery, stringy, and dry. It causes shrinkage of the protein with loss of water from the muscle fibers.
10. Tender cuts of meat can be cooked by dry methods: broiling, pan broiling, frying, roasting, and stir-frying.
11. The less tender cuts of meat are cooked by moist heat methods: braising and stewing.

12. Fully grown chickens are labeled as mature chicken, old chicken, hen, stewing chicken, or fowl. Young chickens, in turn, are labeled as broiler; fryer; broiler-fryer; roaster, capon (castrated male chicken), and Rock Cornish hen.
13. As a rule, it is more economical to buy the whole body chicken and cut it as desired.
14. Young poultry has a low content of connective tissue and is cooked satisfactorily by dry heat methods.
15. Mature birds require a longer, moist heat cookery to make them tender.
16. Whole roasted chicken and turkey should be cooked to an internal temperature of 180°F.

I. TO SHOW THE EFFECTS OF THE DEGREE OF DONENESS AS MEASURED BY THE INTERNAL TEMPERATURE AND THE EFFECTS OF ROASTING TEMPERATURE ON ROASTING TIME, COOKING LOSSES, AND COLOR AND JUICINESS OF GROUND MEAT PATTIES

1. Weigh out four 100 g portions of ground beef.
2. Shape each portion into a round patty of even thickness.
3. To measure the internal temperature, insert short-stemmed or right-angled mercury-filled thermometers so that the bulbs are in the center of the patties. Quick read thermometer could also be used in place of the other thermometers. Timing is important when the patties are cooking in order to monitor end-point temperatures.
4. Put the meat on a small rack. To catch the drippings, place the rack in a shallow pan for which the weight has been previously recorded.
5. Put three samples in an oven at 325°F and roast to the following internal temperatures:
   a. Rare: 130°F (55°C)
   b. Medium: 150°F (65°C)
   c. Well done: 160°F (71°C)
6. Put one sample in an oven at 425°F and roast to 160°F (71°C).
7. Note and record the roasting time for each. As soon as the meat comes from the oven, remove the thermometer, take the meat from the rack, and weigh. Weigh pans with drippings and record in Table 14.1.
8. With a sharp knife cut each patty in half. Rank patties in descending order for uniformity of doneness.
9. Rank samples in descending order for juiciness in Table 14.1.

QUESTIONS

1. What difference does the roasting temperature make in the

   a. roasting time?

   b. total cooking losses?

   c. uniformity of doneness?

   d. juiciness?

2. a. Account for the differences observed in the juiciness of the different samples.

   b. What effect would the type of ground beef, such as sirloin, chuck, or round, have on the results?

3. What are the present standards set for the safe preparation of ground meat?

Table 14.1 QUALITY EVALUATION OF COOKED GROUND BEEF									
Roasting Temperature	Internal Temperature	Cooking Time	Raw Weight	Cooked Weight	% Cooking Loss*	Pan Weight	Drip Weight	Water Loss	Uniform Doneness
325°F	130° (55°C)								
325°F	150°F (65°C)								
325°F	160°F (71°C)								
425°F	160°F (71°C)								

$$*\%\text{Cooking loss} = \frac{\text{Raw weight(g)} - \text{Cooked weight(g)}}{\text{Raw weight (g)}} \times 100$$

II. TO SHOW THE EFFECT OF HEAT AND TREATMENT ON THE COOKING OF LESS TENDER CUTS OF BEEF

A. SWISS STEAK

8 oz. boneless round steak
Dash black pepper
1/4 teaspoon salt
1 tablespoon vegetable oil

1 small yellow onion, diced
1 can (14 oz.) whole or diced tomatoes with liquid
2 teaspoons all-purpose flour
1/4 cup water

1. Trim any excess fat from meat.
2. Season steak with salt or pepper.
3. Brown steak in the hot oil in a medium-sized skillet (the skillet should have a lid that fits snugly) until brown on both sides.
4. Remove meat from the pan; add onions and cook the onions until translucent. Return the steak back to the skillet.
5. Pour the tomatoes over the meat and onions. Bring to a boil. Lower heat, cover pan, and allow the meat to simmer for 45 minutes. Keep checking periodically and add more water as needed to keep the meat and onions from burning.
6. Combine the flour and water. Stir this into the meat mixture; bring to a boil.
7. Evaluate and record observations in Table 14.2 for each treatment.

Treatments

1. *Control:* Follow directions for Swiss Steak.
2. *Pounding:* Pound meat with mallet to 1/4 inch. Proceed with Swiss Steak recipe.
3. *Cutting:* Make diagonal slices on both sides of the meat, not cutting through. Proceed with Swiss Steak recipe.
4. *Enzyme:* Sprinkle both sides of meat with commercial enzyme mixture. Proceed with Swiss Steak recipe.
5. *Marinade:* Marinate meat in 1/4 cup apple cider vinegar, 2 tablespoons soy sauce, 1/4 teaspoon onion powder, and 1/4 teaspoon salt for 15 minutes. Proceed with Swiss Steak recipe.

Table 14.2 EVALUATION OF SWISS STEAK			
Treatment	Tenderness	Juiciness	Appearance of Grain
Control			
Pounding			
Cutting			
Enzyme			
Marinade			

QUESTIONS

1. What procedure yielded the most tender cooked meat?

2. What procedure yielded the juiciest cooked meat?

3. What do you call this cooking procedure for meat?

4. How does it differ from stewing?

5. Which treatment would have required a longer cooking time?

6. In the marinade, which ingredient aided in tenderizing the meat?

7. Identify the enzyme used in the commercial meat tenderizer, and describe the limitations of proteolytic enzymes in increasing the tenderness of meat.

III. TO LEARN TO APPLY DIFFERENT COOKING TECHNIQUES FOR DIFFERENT MEAT CUTS OR VARIETIES

A. BEEF

1. Stir-Fry Beef

Marinade
4 tablespoons low sodium soy sauce
1 tablespoon granulated sugar
2 tablespoons sherry

1/2–3/4 teaspoon ground ginger
2 cloves garlic, minced
1 pound sirloin steak, partially frozen
1 red bell pepper, cut into thin strips

1 cup mushrooms, sliced thin
1/2 head broccoli, cut into flowerets
1 small yellow onion, cut into 1/8 inch
2 tablespoons peanut oil, divided

1 teaspoon sesame seed oil
1 can (8 oz.) sliced water chestnuts, drained
1/2 cup cold water
1 tablespoon cornstarch

1. Prepare marinade by mixing in a medium-sized bowl, the soy sauce, sugar, sherry, ground ginger, and garlic.
2. Slice meat into thin strips. Having the meat partially frozen will allow you to thinly slice the meat. Place meat strips in marinade and marinate for at least 30 minutes.
3. Prepare vegetables. Slice pepper into thin strips. Slice mushrooms 1/8–1/4 inch thickness. Prepare broccoli into flowerets; cut onion into 1/8 inch.
4. Heat 1 tablespoon peanut oil and sesame oil in a 10-inch skillet.
5. Add vegetables: pepper, broccoli, and onion. Stir-fry for 5 minutes or until vegetables are crisp-tender. Remove vegetables and place them on a plate; cover.
6. Drain marinade from meat. Add 1 tablespoon peanut oil to skillet. Add meat and stir-fry until no longer pink. Add reserved marinade, stir-fried vegetables, and water chestnuts to the skillet and cook until all ingredients are heated throughout.
7. Combine 1/2 cup water and cornstarch. Add to skillet; stir constantly. Cook until mixture thickens and comes to a boil.
8. Serve over cooked rice.

2. Beef Stuffed Pepper

4 large green peppers
1 pound ground beef
1 medium onion, chopped
$2\frac{1}{2}$ cups canned marinara sauce, divided
1 cup cooked rice

1 can ($8\frac{3}{3}$ oz.) whole kernel corn, drained
2 teaspoons chili powder
1/2 teaspoon salt
1/2 cup (2 oz.) low-fat Cheddar cheese, shredded

1. Preheat oven to 350°F. Cut off tops of green peppers; remove centers and discard. Cook peppers for 5 minutes in boiling water; drain peppers and set aside.
2. Cook ground beef and onion in a large skillet until meat is browned, stirring to crumble meat; drain well.
3. Stir in 1 cup of marinara sauce and next four ingredients.
4. Stuff peppers with meat mixture, and place in a greased baking dish. Pour in balance of marinara sauce so it surrounds the peppers.
5. Bake in preheated oven for 25 minutes. Sprinkle tops of peppers with Cheddar cheese; bake an additional 5 minutes.

3. Sloppy Joes

2 teaspoons vegetable oil
1 medium onion, coarsely chopped
1 green bell pepper, diced
3 cloves garlic, minced
1 pound lean chopped beef
1/2 cup ketchup
3/4 cup chili sauce
1 tablespoon tomato paste

1/3 cup water
$1\frac{1}{2}$ teaspoons Worcestershire sauce
1 tablespoon chili powder
1 tablespoon cumin
1/2 teaspoon salt
1/4 teaspoon ground black pepper
3 drops hot sauce
4 hamburger buns, split

1. In a large skillet heat oil. Add onion, bell pepper, and garlic.
2. Cook over medium heat until pepper and onion are soft.
3. Add chopped beef. Break up meat until it is crumbly and browned.
4. Add remaining ingredients, except for the hamburger buns.
5. Stir until thoroughly combined and mixture comes to a boil. Lower heat and cover. Cook for 15 minutes.
6. Split hamburger buns in half. Serve hamburger mixture over the buns.

4. Beefy Bean 'N Biscuit Casserole

1½ pounds chopped beef
1/2 cup chopped onion
1/2 cup chopped green pepper
1 cup chopped celery
1 tablespoon light brown sugar, firmly packed
1 teaspoon garlic salt
1/2 teaspoon ground black pepper
1 teaspoon paprika
3/4 cup water
1 tablespoon apple cider vinegar
1/4 teaspoon Tabasco sauce

1/2 teaspoon Worcestershire sauce
16 oz. can pork and beans.

Biscuit Topping
1½ cups all-purpose flour
2 teaspoons double-acting baking powder
1/2 teaspoon salt
1/4 cup butter or margarine
1/2 cup milk
10–12 (1/2 inch) cubes mild Cheddar cheese

1. Heat oven to 400°F. In a 10-inch skillet, brown beef, onion, green pepper, and celery; drain off any excess fat. Add remaining ingredients except the Biscuit Topping; simmer while preparing the biscuits.
2. To prepare the biscuit topping sift together the flour, salt, and baking powder.
3. Cut in the butter until the consistency of cornmeal. Add milk; stir with a fork just until a soft dough forms.
4. Turn dough onto a lightly floured surface; knead gently 10–12 times. Roll out to 1/2 inch thickness; cut with a 2-inch floured cutter.
5. Press cheese cube in the center of each biscuit. Fold dough over cheese; seal well.
6. Pour meat mixture into a greased shallow 2-quart casserole dish. Arrange biscuits over hot meat mixture. Bake for 20–25 minutes or until biscuits are brown. Serve immediately.

5. Muffin Meatloaves

1 tablespoon olive oil
1 cup finely chopped onion
1/2 cup finely chopped carrot
1 teaspoon dried oregano
2 garlic cloves, minced
1 cup ketchup, divided
1 pound ground sirloin or round
1/2 pound ground veal

1 cup finely crushed cracker crumbs (about 20)
2 tablespoons brown mustard
1 teaspoon Worcestershire sauce
1/2 teaspoon salt
1/4 teaspoon ground pepper
2 large eggs
Vegetable spray

1. Preheat oven to 350°F.
2. Heat the olive oil in a large skillet over medium heat. Add chopped onion, chopped carrot, oregano, and minced garlic; sauté for 5 minutes and cool.
3. Combine onion mixture, 1/2 cup ketchup, and the remaining ingredients except the cooking spray in a large mixing bowl. When mixing, use a fork to combine the ingredients (overmixing will toughen the meatloaves).
4. Evenly spoon the meat loaf mixture into 12 muffin cups that have been coated with the vegetable spray. Top each with 2 teaspoons of ketchup. Bake at 350°F for 25 minutes or until a thermometer registers 160°F. Let it stand for 5 minutes before serving.

B. PORK

1. Barbequed Pork on a Bun

1 pork tenderloin (12 oz.)
1/4 teaspoon salt
1/8 teaspoon ground black pepper
1 cup finely chopped onion
2 tablespoons vegetable oil
1 (8 oz.) can tomato sauce

1/4 cup apple cider vinegar
2 tablespoons dark brown sugar, firmly packed
1 teaspoon dried mustard
1/2–3/4 teaspoon crushed red pepper
4 soft hamburger buns

1. Preheat oven to 400°F. Sprinkle tenderloin with salt and pepper. Place meat on a shallow rack in a baking pan. Roast for 30–35 minutes or when an internal temperature reads 160°F on an instant read thermometer.
2. Let the roast cool. Cut the tenderloin into 1/2 inch lengths, then shred with a fork or pull apart with fingers. Set this aside, covered.
3. In a medium skillet, place oil and onion. Cook over medium heat until onion is softened and translucent, about 5–8 minutes.
4. Stir in tomato sauce, cider vinegar, brown sugar, dried mustard, and red pepper; stir to combine.
5. Bring to a boil, stirring constantly. Lower heat, cover, and simmer for 5–8 minutes, or until slightly thickened but still juicy.
6. Stir in shredded pork. Heat through.
7. Serve hot on buns.

2. Pork Lo Mein

Sauce
3 tablespoons hoisin sauce
3 tablespoons low sodium soy sauce
2 tablespoons water
1/2–1 teaspoon hot sauce
8 oz. thin spaghetti or angel hair spaghetti
1/4 cup peanut oil, divided
2 large eggs, beaten
12 oz. pork tenderloin, thinly sliced

1/2 teaspoon salt
1/4 teaspoon ground black pepper
2 teaspoons ground coriander
1 teaspoon ground fresh ginger root
4 cloves garlic, minced
6 scallions cut diagonally
1 red bell pepper, cut into quarters, seeded, then sliced
1 small can sliced water chestnuts

1. Mix together sauce ingredients and reserve.
2. Cook pasta in salted water according to package directions; set aside.
3. While pasta is cooking, heat a tablespoon of peanut oil in a large skillet over high heat. Add beaten eggs and scramble them to light golden brown, remove and reserve.
4. Season the pork with salt, pepper, and coriander. Heat the remaining peanut oil, then add sliced pork and stir-fry for 4 minutes.
5. Push meat to the side and add ginger, garlic, and the vegetables. Stir-fry the vegetables for 2 minutes.
6. Add cooked pasta and eggs back to the skillet.
7. Pour sauce over the spaghetti and toss to combine. Turn off heat. Toss for 30 seconds and let the liquids absorb.

3. Sweet and Sour Pork

(Reprinted from *BettyCrocker.com* with the permission of General Mills, Inc.)

2 pounds pork boneless top loin
Vegetable oil
1/2 cup all-purpose flour
1/4 cup cornstarch
1/2 cup cold water
1/2 teaspoon salt
1 egg
1 can (20 oz.) pineapple chunks in syrup, drained and syrup reserved
1/2 cup packed light brown sugar

1/2 cup white vinegar
1/2 teaspoon salt
2 teaspoons soy sauce
2 medium carrots, cut into thin diagonal slices
1 garlic clove, finely chopped
2 tablespoons cornstarch
2 tablespoons cold water
1 medium green pepper, cut into 3/4 inch pieces
8 cups hot cooked rice

1. Trim excess fat from pork. Cut pork into 3/4 inch pieces.
2. Heat 1 inch oil in a deep fryer or Dutch oven at 360°F.
3. Beat flour, 1/4 cup cornstarch, 1/2 cup cold water, 1/2 teaspoon salt, and the egg in a large bowl with a hand beater until smooth. Stir pork into batter until well coated.
4. Add pork pieces one at a time to the oil. Fry about 20 pieces at a time for about 5 minutes, turning 2–3 times, until golden brown. Drain on paper towels; keep warm.

5. Add enough water to reserved pineapple syrup to measure 1 cup. Heat syrup mixture, brown sugar, vinegar, 1/2 teaspoon salt, the soy sauce, carrots, and garlic to boiling in Dutch oven; reduce heat to low.
6. Cover and simmer about 6 minutes or until carrots are crisp-tender. Mix 2 tablespoons cornstarch and 2 tablespoons cold water; stir into sauce.
7. Add pork, pineapple, and bell pepper. Heat to boiling, stirring constantly. Boil and stir for 1 minute. Serve with cooked rice.

C. LAMB

1. Lamb Curry

1 pound boneless lamb shoulder
1 teaspoon salt
1 garlic clove, minced
1 large onion
1 cup celery, diced

2 tablespoons vegetable oil
$1/2$–1 teaspoon curry powder
1 medium Granny Smith apple
Cooked rice

1. Cut the meat into 1 inch cubes. Trim away excess fat while cubing.
2. Place the cubed lamb in a saucepan; barely cover with water and add salt; cover and simmer for approximately 1 hour. Remove meat from saucepan.
3. Cut onion into 1/4 inch thick slices. Dice celery. Clean garlic and impale the garlic clove with a toothpick. Add these vegetables to the saucepan with the vegetable oil. Cook for 5 minutes. Remove the garlic clove.
4. While the onion–celery mixture is cooking, wash, pare, core, and finely chop the apple. Add to the onion–celery mixture; add the curry powder (add the small amount first—then add as desired). Add the lamb with whatever liquid remains on the lamb. Simmer uncovered for approximately 20 minutes (mixture may be covered if there is no excess water on the lamb).
5. Serve over cooked rice.

D. VARIETY (ORGAN) MEAT

1. Sautéed Calf's Liver

4 slices turkey bacon
2 tablespoons olive oil
1/3 cup all-purpose flour
1/2 teaspoon salt
1/4 teaspoon ground black pepper
3/4 pound beef calves' liver

2 tablespoons butter or margarine, melted
1 small onion chopped
2 tablespoons fresh parsley, chopped
1 tablespoon lemon juice
Pinch nutmeg
Pinch savory

1. Add olive oil to skillet; add turkey bacon and cook over medium heat until crisp; remove bacon. Drain the bacon on a paper towel; crumble and reserve.
2. Combine flour, salt, and pepper; dredge liver in flour and brown in the reserved olive oil plus one tablespoon butter. Remove liver to a warm platter and top with crumble bacon.
3. Sauté onion and parsley in remaining butter. Stir in lemon juice, nutmeg, and savory; pour over liver.

QUESTIONS

1. What cuts of meat are suitable for roasting?

2. What is the most accurate way of measuring degree of doneness in meat?

150

3. To what end-point temperature should pork be cooked to and why?

4. What is meat by variety meat and how should it be cooked?

IV. TO OBSERVE AND LEARN HOW TO CUT UP OR DISJOINT A WHOLE CHICKEN

A. DEMONSTRATE HOW TO CUT UP OR DISJOINT CHICKEN

1. Cut the skin parallel to the fold between the thigh and the body cavity of the bird.
2. Grasp the body of the chicken in the left hand and the leg and thigh in the right hand, and bend the latter back until the joint snaps. Cut the thigh muscle from the back as close to the back as possible.
3. To separate the leg from the thigh, bend to locate the joint. Then cut at the joint between leg and thigh. When halfway through, reverse direction and cut.
4. Detach wing from the body at the joint.
5. Separate back from ribs.
6. Cut sternum or breast bone from ribs.

V. TO LEARN VARIOUS METHODS FOR CHICKEN PREPARATION

A. CHICKEN AND DUMPLINGS

1/2 cup all-purpose flour	1 cup water
1½ teaspoons salt	Pinch of thyme and rosemary
2 teaspoons paprika	1 onion, chopped
1/4 teaspoon ground pepper	3 tablespoons all-purpose flour
3 pound broiler-fryer, cut up	Milk
Shortening or vegetable oil	Rosemary and Chive Dumplings (below)

1. Mix 1/2 cup flour, salt, paprika, and pepper. Coat the chicken pieces with flour mixture.
2. Heat a thin layer of shortening in a 12-inch skillet or Dutch oven until hot.
3. Cook chicken in shortening until brown on all sides. Drain fat from skillet; reserve. Add water, onion, and pinch of rosemary and thyme.
4. Cover and cook over low heat for 45 minutes until chicken is tender.
5. Remove chicken from skillet. Remove chicken from the bone. Keep chicken in large meaty chunks. Heat 3 tablespoons of reserved fat in the skillet. Stir in 3 tablespoons of flour. Cook over low heat until mixture is smooth and bubbly.
6. Add enough milk to the reserved liquid to measure 3 cups; pour into skillet. Heat to boiling, stirring constantly. Boil and stir for 1 minute. Return chicken to the gravy.
7. Prepare Rosemary and Chive Dumplings; drop by spoonfuls onto hot chicken. Cook uncovered for 10 minutes; cover and cook for 20 minutes longer.

Rosemary and Chive Dumplings

3 tablespoons chopped chives	3/4 teaspoon salt
1 teaspoon dried rosemary, crumbled	3 tablespoons solid shortening
1½ cups all-purpose flour	3/4 cup milk
2 teaspoons double-acting baking powder	

1. Mix together the chives, rosemary, flour, baking powder, and salt. Cut in shortening with pastry blender.
2. Stir in milk, just until ingredients are combined.
3. Drop by spoonfuls onto hot chicken. Proceed as in Step 7 in the above.

B. STIR-FRY CHICKEN

Use Stir-Fry Beef Recipe except:

1. Substitute 1 pound boneless skinless chicken breasts, thinly sliced, for the beef.
2. Substitute 1/2 cup chicken broth for 1/2 cup water.
3. Proceed as the recipe indicates.

C. CHICKEN LO-MEIN

Use Pork Lo-Mein recipe except:

1. Substitute 1 pound of boneless skinless chicken breasts, thinly sliced, for the pork tenderloin.
2. Proceed as the recipe indicates.

D. BAKED LEMON CHICKEN

1 egg white	1/4 teaspoon ground red pepper (cayenne),
1 tablespoon water	if desired
4 boneless and skinless chicken breasts	Cooking spray
(about 1¼ pounds)	Lemon sauce (below)
1/2 cup all-purpose flour	1/2 lemon, cut into thin slices
1 teaspoon baking soda	1 medium green onion, chopped

1. Pound chicken with a mallet between two pieces of plastic wrap to thin slightly. In a medium shallow dish mix egg white and water. Add chicken; turn chicken to coat both sides.
2. Heat oven to 425°F. Spray a 15 × 10 × 1 inch pan with vegetable cooking spray. In a resealable plastic food storage bag mix flour, baking soda, and red pepper. Remove chicken from egg white mixture. Add one chicken breast at a time to the flour mixture. Seal bag; shake to coat chicken. Place chicken on the prepared baking sheet. Spray the flour-coated chicken with the vegetable spray for about 5 seconds or until the surface of the chicken appears moist.
3. Bake chicken uncovered for 20–25 minutes or until the juice of the chicken is no longer pink when the thickest part of the chicken is pierced with a knife.
4. Meanwhile, make Lemon Sauce. Let chicken stand for 5 minutes; cut each breast diagonally into five slices. Pour warm Lemon Sauce over chicken. Garnish with lemon slices and green onion. Serve with Yellow Rice (see below)

Lemon Sauce

1/4 cup granulated sugar	2 tablespoons rice vinegar
1/3 cup chicken broth	1/4 teaspoon salt
1 teaspoon grated lemon rind	1 garlic clove, minced
3 tablespoons fresh lemon juice	2 teaspoons cornstarch
3 tablespoons light corn syrup	2 tablespoons water

1. In a 1-quart saucepan heat all the ingredients except for the cornstarch and cold water.
2. Bring mixture to a boil, stirring occasionally.
3. In a custard cup mix cornstarch and water; stir into sauce. Cook and stir until mixture comes to a boil.

Yellow Rice
Prepare Rice Pilaf (Laboratory 3) except use chicken broth for the liquid and add 1/8 teaspoon turmeric to the sautéed rice. Cook the rice as directed in the recipe. Serve chicken over the rice.

E. OVEN BARBECUED CHICKEN

3 pounds of chicken pieces	1/4 teaspoon ground black pepper
1/2 cup all-purpose flour	1/4 cup margarine, melted
1 teaspoon salt	Texas Barbecue Sauce (below)

1. Preheat oven to 425°F. Spray a 13 × 9 × 2 inch pan with vegetable spray; set aside.
2. Mix together the flour, salt, and pepper. Toss chicken pieces in the seasoned flour to coat.
3. Place coated chicken pieces fat side down in pan; spoon melted margarine evenly over the chicken pieces. Bake for 30 minutes.
4. While chicken is cooking, prepare Texas Barbecue Sauce (below).
5. After the 30-minute baking time for the chicken, turn chicken pieces over. Spoon half of the Texas Barbecue Sauce over the cooked chicken.
6. Baste chicken with the balance of the barbecue sauce. Bake for another 30 minutes or until tender.

Texas Barbecue Sauce

1 tablespoon granulated sugar
1 tablespoon dark brown sugar, packed
1 tablespoon paprika
1 teaspoon salt
1 teaspoon dry mustard
1/4 teaspoon chili powder
1/2 medium onion, chopped

1 cup tomato sauce
1/2 cup water
1/4 cup ketchup
$1\frac{1}{2}$ tablespoons Worcestershire sauce
1/4 cup apple cider vinegar
1/8 teaspoon cayenne pepper

1. Mix all ingredients in a saucepan. Simmer uncovered for 15 minutes.

F. CHILI CON CARNE

1 tablespoon olive oil
$1\frac{1}{4}$ pounds ground turkey or chicken
1 large yellow onion, finely chopped
1/2 medium green pepper, cored, seeded, chopped
1/2 medium red pepper, cored, seeded, chopped
2 stalks celery, chopped
2 garlic cloves, minced
1 tablespoon chili powder
2 teaspoons ground cumin

1 can ($10\frac{1}{4}$ oz.) tomato puree
1 cup hot water plus 1 beef bouillon cube
2 tablespoons tomato paste
1/4 teaspoon crushed red pepper flakes
1 teaspoon ground coriander
1 teaspoon dried oregano, crumbled
1 teaspoon dried basil, crumbled
1 bay leaf
1 can (15 oz.) red kidney beans, undrained

1. Add olive oil to a nonstick 10-inch skillet. Heat pan over medium heat. Add the ground turkey or chicken; cook; stir often until the meat is no longer pink.
2. Add the onion, green and red peppers, celery, garlic, chili powder, and cumin; cook until vegetables are soft—about 5 minutes. Add the tomato puree, beef broth, tomato paste, red pepper flakes, coriander, oregano, basil, and bay leaf. Simmer, partially covered, for 20 minutes, stirring occasionally.
3. Add the kidney beans and simmer, partially covered, for 5 minutes longer, stirring occasionally. Discard the bay leaf. Serve with cooked rice or with Buttermilk Cornbread (recipe below).

Buttermilk Cornbread

1 cup all-purpose flour
1/2 cup yellow cornmeal
1/2 cup white cornmeal
2–4 tablespoons granulated sugar
1 tablespoon baking powder

1/4 teaspoon salt
1/4 teaspoon baking soda
2 eggs
1 cup buttermilk
1/4 cup vegetable oil

1. Preheat oven to 425°F. Grease a 9 × 9 × 2 inch pan. Set aside.
2. In a medium-sized mixing bowl sift together flour, cornmeal, and desired level of sugar, baking powder, salt, and baking soda.
3. In another small bowl mix together the eggs, buttermilk, and oil. Add the liquid mixture to the dry ingredients and stir just until moistened (batter should be lumpy).
4. Pour batter into prepared pan. Bake in preheated oven for 20–25 minutes or until brown. Serve warm.

G. CHICKEN THIGHS PARADISE

Ginger–Honey Glaze

1/2 cup orange juice
1/2 teaspoon grated orange rind
1/4 cup soy sauce
1 tablespoon fresh ginger, pared and chopped
1 tablespoon granulated sugar
1 tablespoon honey
Butter, margarine, or vegetable oil (for basting chicken)

1 clove garlic, minced
1/4 cup chopped green onion
$1\frac{1}{2}$ teaspoons cornstarch dissolved in
 2 tablespoons water
1 teaspoon white vinegar
$1\frac{1}{2}$ pounds skinless chicken thighs

1. In a saucepan stir together the Ginger–Honey Glaze ingredients.
2. Cook and stir until bubbly. Continue cooking, stirring constantly, for 2 minutes.
3. Preheat broiler. Line broiler pan with aluminum foil. Spray rack with nonstick vegetable oil cooking spray.
4. Place chicken, meat side down, on broiler pan.
5. Broil chicken 5 inches from heat for 7 minutes. Broil about 15–20 minutes longer, brushing occasionally with butter, margarine, or oil.
6. Turn chicken and broil for 5–15 minutes more or until tender and no longer pink.
7. During the last 5 minutes of cooking, brush the chicken with Ginger–Honey Glaze. Heat the remaining glaze; serve with chicken.

QUESTIONS

1. Why can chicken be cooked by various cooking methods?

2. Give instructions for using a thermometer when roasting poultry.

3. Briefly describe what quality attributes to look for when buying poultry.

4. What are the effects of overcooking poultry?

5. How should be poultry be handled when it is brought home from the supermarket?

6. Which microorganisms are associated with raw poultry?

7. What are the nutritional consequences in consuming poultry versus beef?

LABORATORY 15

Fish and Seafood Cookery

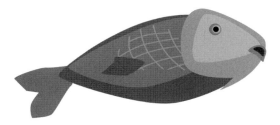

LABORATORY 15
Fish and Seafood Cookery

Because of a more healthy eating style, people are urged to consume more seafood in their daily diet. Selection and preparation techniques are required to ensure an acceptable product. This laboratory exercise will introduce the student to the proper techniques in selection of quality seafood, and how to preserve these qualities through various preparation techniques.

VOCABULARY

crustaceans	finfish	lean fish
drawn fish	fillet	mollusks
dressed fish	kippered	roe
fatty fish		

OBJECTIVES

1. To learn to identify the certain market forms of seafood.
2. To learn how to identify freshness characteristics of seafood.
3. To learn how to prepare seafood to maintain quality and nutritional value.

PRINCIPLES

1. Two major categories for the classification of fish are:
    a. vertebrate fish with fins,
    b. shellfish or invertebrates.
2. Shellfish are of two types:
    a. mollusks are soft structured and are either partially or wholly enclosed in a hard shell.
    b. crustaceans are covered with a crustlike shell and have segmented bodies.
3. Examples of mollusks are oysters, clams, abalone, scallops, and mussels. Examples of crustaceans are lobster, crab, shrimp, and crayfish.
4. Finfish can be lean or fatty. If fatty, the oil content is greater than 5%, while lean finfish have an oil content of less than 5%. Fatty finfish will be higher in calories, but will have a stronger flavor. The oils contained in the fish are the long-chain omega-3 and -6 fatty acids that are desired in the diet.
5. High-fat fish are lake trout, pompano, salmon, white fish, and catfish.
6. Lean fish are cod, flounder, grouper, haddock, halibut, ocean perch, orange roughy, pike, red snapper, sea bass, sole, and swordfish.
7. There are ways to check for freshness in fish:
    a. eyes are clear and bright.
    b. gills are bright red.
    c. flesh is firm and pliable.
    d. odor is mild with no offensive smell.
8. According to food experts, a variety of different fish, whether they come from cold or warm water, at times their gills and eyes will not always follow the above criteria. However, the aroma and texture are primary objectives in fish selection.
9. If fish is frozen, it should be packaged tightly wrapped and sealed. It should be solidly frozen and free of ice crystals. Ice crystals indicate fish was thawed and frozen.
10. The market forms of fish are (Fig. 15.1):
    a. Whole.
    b. Drawn: entrails are removed.
    c. Dressed: head, scales, and fins are removed from a drawn fish.
    d. Steaks: taken from a large dressed fish; they are cross-section slices.
    e. Fillets: sides of the fish are cut lengthwise away from the backbone; they are practically boneless.
11. When cooking fish, choose from baking, broiling, grilling, frying, steaming, and poaching.
12. Dry heat is considered better for cooking fatty fish. Lean fish remains moist when cooked by moist heat methods.

13. Fish does not contain connective tissue that is found in meat and poultry, and overcooking is a common problem. This will dry and toughen the fish, and will, also, destroy the flavor.
14. Check occasionally for doneness while cooking. Pierce the thickest part of the fish. Most fish will flake easily when done, lose their translucency, and become opaque.
15. Fish can be cooked in the microwave. Microwave HIGH power to quickly seal the juices and flavor. Arrange the thicker portion to the outside of the dish so they will cook without drying out the thinner areas.

Whole or round fish	Dressed or pan-dressed fish	Steaks	Drawn fish	Single fillet

FIG. 15.1: Market forms of fish.

I. HOW TO PREPARE FISH IN A VARIETY OF WAYS

A. BROILED FISH STEAKS

3 fish steaks, such as salmon or swordfish, 1 inch thick
1/2 teaspoon salt

Dash of ground black pepper
2 tablespoons butter or margarine, melted

1. Sprinkle both sides of the fish with salt and pepper; brush with half of butter.
2. Set oven control to broil. Broil with top of fish about 4 inches from the heat until lightly brown, about 6 minutes. Brush fish with butter.
3. Turn the fish over carefully; brush with butter; broil until fish flakes very easily with the fork and is opaque in the center, for 4–6 minutes longer.

B. OVEN FRIED FISH

1 pound fish fillets
2 tablespoons cornmeal
2 tablespoons dried bread crumbs
1 tablespoon grated Parmesan cheese
1/4 teaspoon salt
1/4 teaspoon paprika

1/8 teaspoon garlic powder
1/8 teaspoon dried dill weed
Dash ground black pepper
2 egg whites, slightly beaten with 1 tablespoon water
3 tablespoons olive oil (or melted butter or margarine)

1. Move oven rack to position slightly above the middle of the oven. Preheat oven to 500°F.
2. Cut fish fillets into 2 × 1½ inch pieces.
3. Mix cornmeal, bread crumbs, Parmesan cheese, salt, paprika, garlic powder, dill weed, and pepper. Dip fillet in egg white mixture; coat with cornmeal mixture.
4. Place coated fillet pieces on a greased 13 × 9 × 2 inch pan. Dribble the olive oil evenly over the coated pieces of fish.
5. Bake uncovered until fish flakes very easily with a fork, about 10 minutes.

C. PAN-FRIED FISH

1 pound fish fillets
1/2 teaspoon salt
Dash of ground black pepper
1 egg

1 tablespoon water
1/2 cup all-purpose flour or cornmeal
Shortening (part butter or margarine)

1. If fish fillets are large, cut into serving pieces. Sprinkle both sides of fish with salt and pepper.
2. Beat egg and water until blended. Dip fish in egg, then in flour or cornmeal to coat.
3. Heat shortening (1/8 inch) in skillet until hot. Fry fish in shortening over moderate heat, turning fish carefully, until brown on all sides (about 10 minutes).

157

D. STUFFED–BAKED FISH

Stuffing

3 tablespoons butter	3 cups fresh bread cubes
3/4 cup chopped celery	1/2 teaspoon salt
1/2 cup minced onion (or 2 tablespoons dried minced onion + 2 tablespoons water)	1/2 teaspoon savory
	Dash ground black pepper

1. Over moderate heat melt butter in a medium-sized skillet. Add onion and celery and cook the vegetables until they are softened. **If using dried minced onion, just cook celery alone. Combine rehydrated onion and bread cubes with the cooked celery.**
2. Combine the bread cubes, salt, savory, and pepper in a large bowl. Add the cooked vegetables and mix thoroughly. If the bread seems dry, add some water, 1 tablespoon at a time.
3. Set the stuffing aside while preparing the fish.

4 large fish fillets, about 1/2 inch thick	Old Bay Seasoning
Salt	2 tablespoons butter
Ground black pepper	

1. Preheat oven to 400°F. Lightly grease a 13 × 9 × 2 inch pan with vegetable spray.
2. Wash fillets under cold running water and dry thoroughly between absorbent paper towels.
3. Sprinkle salt and pepper over the 4 fillets. Place 2 fillets side by side in the pan. Place the stuffing over the 2 fillets. Layer the remaining 2 fillets over the stuffing. Sprinkle Old Bay Seasoning over the top fillets. Dot butter over the fillets.
4. Place pan in the oven and bake. Allow 20–25 minutes for the fish to cook. The fish will flake with a fork when ready.

Variation

3 tablespoons of white wine plus 2 tablespoons of fresh lemon juice can be poured over the fish before being placed in the oven.

QUESTIONS

1. Give directions for cooking fish by

   a. broiling.

   b. baking.

   c. frying.

   d. steaming.

   e. microwaving.

2. Explain why it is appropriate to cook fish by either moist or dry heat.

3. a. Describe typical characteristics for high-quality fresh fish.

   b. Suggest appropriate procedures for handling and storing fish. Why are these procedures necessary?

4. Why are cooking times different for fish than for meat?

II. TO LEARN HOW TO PREPARE SHELLFISH

A. SHRIMP

Shrimp, fresh or frozen (thawed) in the shell.

1. Peel off the shell of the fresh or frozen (thawed) shrimp.
2. With a sharp knife, cut along the outside of the center back only deep enough to expose the sand vein.
3. Remove the sand vein, which may vary in color from light tan to black depending on the contents.
4. Wash the shrimp in running, cold tap water. Hold in clean cold water until ready to cook.
5. Add 3/4 teaspoon salt to each cup of water to cook the shrimp.
6. Bring the water to a boil.
7. Add the cleaned shrimp to the boiling water. Reduce heat.
8. Cook the shrimp at a simmering temperature, 185–200°F for 5 minutes* or until the shrimp becomes opaque and some portions of the outer surface become a light coral pink.
9. Drain. Put the shrimp into ice water to chill if they are to be served cold.

*To determine the effect of overcooking, leave one or two shrimp in the cooking water and boil (212°F) these shrimp for 10 minutes. Observe the shrinkage during the cooking period. Compare with the simmered shrimp for tenderness.

1. Shrimp Scampi

2 pounds medium fresh shrimp	1/4 cup white wine or vermouth
1/4 cup flat leaf parsley	4 tablespoons fresh lemon juice
4 cloves garlic, crushed	1 teaspoon salt
2 tablespoons butter	1/4 teaspoon ground black pepper
2 tablespoons olive oil	

1. Peel and devein shrimp.
2. Sauté parsley and garlic in butter and olive oil until garlic is lightly browned.
3. Reduce heat to low; add shrimp. Cook, stirring frequently, until shrimp is pink, about 2–5 minutes.
4. Remove shrimp with a slotted spoon to a serving dish; keep warm.
5. Add remaining ingredients to butter mixture; simmer for 2 minutes; pour mixture over shrimp.

2. Spicy Shrimp Creole

1½ pounds fresh shrimp, unpeeled	1/2 cup celery, chopped
1 small onion, chopped	2 medium garlic cloves, minced
1 small green pepper, chopped	2 tablespoons butter or margarine, melted

2 slices turkey bacon, cut into 1/4 inch pieces
1 can (16 oz.) whole tomatoes, undrained and chopped
1 can (8 oz.) tomato sauce
1 tablespoon fresh lemon juice
1 tablespoon light brown sugar, firmly packed

2 teaspoons Worcestershire sauce
1/2 teaspoon dried whole thyme
1/8 teaspoon red pepper
1 bay leaf
Hot cooked white rice

1. Peel and devein shrimp.
2. Sauté onion, green pepper, celery, garlic, and turkey bacon in butter in a Dutch oven until tender.
3. Stir in tomatoes, tomato sauce, lemon juice, brown sugar, Worcestershire sauce, thyme, red pepper, and bay leaf.
4. Cook over medium heat, stirring occasionally, about 30 minutes or until desired consistency.
5. Stir in shrimp and simmer over medium heat for 5–10 minutes or until shrimp are cooked (curled and pink). Serve over cooked rice.

3. Szechuan Shrimp

Brined Shrimp
2 pounds fresh (not frozen), deveined shrimp
1½ cups kosher salt

1 cup boiling water
2 cups cold water

1. Mix together kosher salt and boiling water in a large bowl. Add the 2 cups of cold water; mix. **Some of the salt will not dissolve.**
2. Add the shrimp. Let it stand for 20 minutes.
3. Rinse the shrimp thoroughly three or four times to make sure the salt is not remaining on the shrimp.
4. Devein the shrimp, and refrigerate while preparing the Szechuan sauce.

Szechuan Sauce
1/2 cup ketchup
4 tablespoons low sodium soy sauce
2 tablespoons sherry

1 teaspoon sugar
1 tablespoon minced fresh ginger
1/2 teaspoon crushed red pepper flakes

1. In a small bowl combine all the ingredients and set aside.

Completing the Dish
2 tablespoons peanut oil
Brined shrimp (above)

1 tablespoon minced garlic
1 cup sliced green onions
Szechuan sauce (above)

1. In a large skillet, heat 1 tablespoon of peanut oil over medium high heat. Add shrimp; stir-fry until pink and curled, about 3 minutes. Remove shrimp with a slotted spoon to a bowl.
2. Add remaining tablespoon peanut oil to the skillet; add garlic and scallions and stir-fry for 1 minute. Add Szechuan sauce, cook, stirring constantly until bubbly, about 30 seconds.
3. Add shrimp and stir to combine with sauce. Heat for 1 minute. Serve over cooked rice.

4. Stir-Fried Shrimp with Vegetables

1 pound medium fresh shrimp, unpeeled
1/2 teaspoon salt
1 teaspoon sesame oil or vegetable oil
1 cup water
3 tablespoons oyster sauce
2 teaspoons cornstarch
1 teaspoon chicken bouillon granules
1/4 cup peanut oil or vegetable oil
2 cloves garlic, crushed
2 teaspoons fresh ginger root, grated

1 medium sweet red pepper, cored, seeded, minced
1 can (4 oz.) water chestnuts, drained, sliced thin
2 cups, diced celery
1/2 pound fresh mushrooms, wiped clean, sliced thin
8 scallions, trimmed, minced
1/2 pound fresh snow pea pods or 1 package (6 oz.) frozen snow pea pods, thawed
2 teaspoons rice wine or white wine

1. Peel and devein the shrimp. Sprinkle shrimp with salt and toss with sesame seed oil.
2. Combine water, oyster sauce, cornstarch, and bouillon granules; stir well. Set mixture aside.
3. Pour peanut oil around the top of a preheated wok, coating sides. Allow to heat for 1 minute.
4. Add garlic and ginger root and stir-fry for 30 seconds. Add shrimp and stir-fry for 1½ minutes. Remove shrimp with a slotted spoon and drain on paper towels.
5. Add red pepper, water chestnuts, mushrooms, celery, and scallions to the wok. Stir-fry for 2 minutes. Add pea pods; stir-fry for 30 seconds. Add broth mixture, stir-fry constantly until slightly thickened.
6. Stir in shrimp and rice wine. Serve immediately over boiled rice or thin noodles.

QUESTIONS

1. What would cause the shrinkage and toughness of shrimp when cooked?

2. a. What purpose does brining serve when preparing shrimp?

   b. Explain how brining can affect the cooking quality of poultry and pork?

3. How do the methods of simmering a piece of fish and a cut of beef differ?

B. SCALLOPS

1. Broiled Ginger Scallops

1 pound scallops
1/4 cup soy sauce
1/2 teaspoon minced garlic
2 tablespoons ginger root, finely chopped
2 tablespoons sherry

2 tablespoons freshly squeezed lemon juice
1 tablespoon peanut oil
1/2 teaspoon sesame oil (optional)
1 tablespoon honey

1. If scallops are large, cut in halves. Arrange scallops in a single layer in an 8 × 8 × 2 cubic inch baking dish.
2. Heat soy sauce to boiling. Add garlic and ginger root; reduce heat. Simmer, uncovered, for 3 minutes. Stir in remaining ingredients; pour over scallops. Cover and refrigerate, stirring occasionally, for 2 hours.
3. Set oven control to broil. Remove scallops from marinade with slotted spoon. Arrange in a single layer on rack in broiler pan.
4. Broil with tops 3–4 inches from heat until opaque in center, about 5 minutes. Brush frequently with marinade.

C. SEAFOOD ANALOGS (CRAB, LOBSTER, OR SCALLOPS)

1. Seafood Newburg (Low Fat)

4 teaspoons all-purpose flour
1½ cups skim milk
2 tablespoons white wine or sherry
1/2 teaspoon salt
1/2 teaspoon dry mustard

4 oz. mushrooms, sliced thin
2 teaspoons butter
3 green onions with tops, sliced thin
Salt and ground black pepper

1 package (8 oz.) seafood analog (crab, lobster, or scallops)
Paprika

2 tablespoons grated Romano or Parmesan cheese
7 oz. cooked pasta (linguini or fettuccine)

1. In a 1½-quart saucepan mix with a wire, whisk the flour and milk. It is important that the flour is thoroughly dissolved.
2. Add the wine, salt, and dry mustard.
3. Cook the mixture over medium heat, stirring constantly until mixture boils, boil and stir for 1 minute.
4. In an 8- or 9-inch skillet, melt butter. Add mushrooms and green onion. Stir-fry over medium heat until mushrooms are softened. Season to taste with salt and pepper.
5. Add cooked mushrooms and seafood analog to the cream sauce. Bring sauce back to a boil.
6. Place sauce in a shallow pan (scallop dish or pie dish). Sprinkle the mixture with Romano cheese and paprika. Place under preheated broiler and broil until cream sauce is bubbly.
7. Remove from heat source and serve immediately over pasta.

D. ACCOMPANYING SAUCES

1. Cocktail Sauce

2 tablespoons ketchup
1 tablespoon fresh lemon juice
1/4 teaspoon salt

1 teaspoon horseradish
1/4 teaspoon Worcestershire sauce
1–2 drops Tabasco

1. Blend together all ingredients.
2. Chill for best flavor.
3. Served as an accompaniment to cooked shrimp.

2. Tartar Sauce

1/3 cup mayonnaise
2 teaspoons sour cucumber pickles, minced
2 teaspoons green olives, minced
1/2 teaspoon capers

1/2 teaspoon green onion, minced
1/2 teaspoon parsley, minced
1/2 teaspoon tarragon vinegar

1. Mix all ingredients together.
2. Serve as a accompaniment to fried fish.

3. Drawn Butter Sauce

3 tablespoons butter or margarine
1½ tablespoons flour
1 cup boiling water

1/4 teaspoon salt
Few grains cayenne pepper

1. Melt 2 tablespoon butter in a 1-quart saucepan.
2. Add flour; stir until blended.
3. Add the boiling water gradually; stir until smooth after each addition of water.
4. Bring to a boil over direct heat with continuous stirring.
5. Stir in the remaining butter just before serving. Add salt and cayenne pepper. Serve hot.

GENERAL QUESTIONS

1. What recommendations should be given when cooking shellfish?

2. How important are sanitary conditions when cooking shellfish?

3. What are the nutritional consequences in eating:

   a. finfish?

   b. shellfish?

4. What is the shelf life of fresh fish? How can it be extended?

LABORATORY 16

Legumes and Tofu

LABORATORY 16
Legumes and Tofu

Legumes are an alternate source of protein as well as complex carbohydrates. Legumes are usually in the dry state; therefore, they contain more protein. Tofu is derived from soybeans. Tofu is known as bean curd which is made by coagulating soymilk. This laboratory will introduce the student to the basic techniques in the preparation of legumes and tofu.

VOCABULARY

firm tofu	nigari	soft tofu
legume	rehydration	soybeans
lentils	silken tofu	tofu
lima bean		

PRINCIPLES

1. Legumes, besides being sources of incomplete protein, are rich sources of complex carbohydrates, including both starch and nondigestible carbohydrate.
2. Legumes should be sorted by removing pebbles and broken or decayed beans. The surface soil should be washed off.
3. Legumes are usually soaked in water overnight prior to cooking.
4. Another alternate and quick method is to heat the soaked legume for 2 minutes to boiling and then soaking in the hot water for 1 hour prior to cooking. This process hastens the rehydration process.
5. Acid interferes with the softening of the legumes during cooking by delaying the softening of the cellulose and by inhibiting the softening of the pectin.
6. If legumes are to be seasoned with an acidic ingredient, such as tomatoes, vinegar, brown sugar, ketchup, molasses, sweet and sour sauce, the ingredient should not be added until the beans are tender.
7. Calcium and magnesium ions which are found in hard water combine with pectic substances in legumes to form insoluble complexes giving the legume a firm, almost woody texture.
8. Addition of 1/8 teaspoon of baking soda per cup of dry beans counteracts the effects of hard water. Excessive amounts of baking soda produce an alkaline medium that produces an off flavor to the legume, and is destructive to thiamine.
9. Tofu, also known as soybean curd, is a soft, cheeselike food made by curdling fresh hot soymilk with a coagulant.
10. Traditionally, the curdling agent used to make tofu is nigari, a compound found in natural ocean water, or calcium sulfate, a naturally occurring mineral.
11. In recipes, tofu acts as a sponge and has the ability to soak up the flavor that is added to it.
12. Three types of tofu are available to the consumer:
    a. *Firm tofu*: dense and solid and holds up well in stir-fry dishes, soups or on the grill. It is also higher in protein, calcium, and fat than the other forms of tofu.
    b. *Soft tofu*: a good choice for recipes that call for blended tofu.
    c. *Silken tofu*: it is made by a slightly different process that results in a creamy, custard-like product.
13. Tofu is rich in high-quality protein. It is also a good source of B vitamins and iron.
14. When the curdling agent used to make tofu is calcium sulfate, the tofu is an excellent source of calcium.
15. Tofu is low in saturated fat and contains no cholesterol.
16. Approximately, 1/4 pound of tofu contains about 40 mg of healthy isoflavones which current research is showing greater health benefits with isoflavone intake.
17. Tofu most commonly is sold in water-filled tubs, vacuum packs, or in aseptic brick packages.
18. Tofu can be frozen, and when defrosted has a pleasant caramel color and a chewy, spongy texture that resembles chopped meat. This texture is able to pick up any marinade used in the recipe.

I. TO LEARN HOW TO IDENTIFY AND PREPARE LEGUMES

A. IDENTIFICATION OF LEGUMES

Selected legumes will be left on a tray. You will identify them. Fill in the table (Table 16.1) provided.

Table 16.1 EVALUATION OF LEGUMES			
Legume	Shape	Color	Description

B. RECIPES WITH LEGUMES

1. Cheesy Lentils

2 cups water
6 oz. dried lentils
1 tablespoon instant granulated chicken
   bouillon
1 medium onion, chopped

1 tablespoon red wine vinegar
1 medium green pepper, chopped
1 medium tomato, seeded and chopped
1 cup grated Cheddar cheese or reduced fat
   variety

1. Heat water and lentils to boiling in a 2-quart saucepan; stir in bouillon.
2. Cover and simmer until lentils are tender, about 30 minutes. Add more water during cooking, if necessary.
3. Stir in onion, vinegar, and green pepper; simmer uncovered for 5 minutes.
4. Stir in tomato. Sprinkle with Cheddar cheese.

2. Italian Lima Beans

2 cups water
1/2 cup dried lima beans
1 teaspoon salt
2 tablespoons olive oil
1 medium onion, finely chopped
1 medium green pepper, chopped

1 cup celery, thinly sliced
1 cup carrots, thinly sliced
1/2 teaspoon dried basil leaves
1 can (8 oz.) tomato sauce
1/4 cup Parmesan or Romano cheese, grated

1. Heat water, beans, and salt in a Dutch oven. Boil for 2 minutes; remove from heat. Cover and let it stand for 1 hour.
2. Add enough water to cover, if necessary. Heat to boiling; reduce heat. Cover; simmer until tender, for $1\frac{1}{4}$–$1\frac{1}{2}$ hours. Add more water during cooking.
3. Preheat oven to 375°F.
4. Drain beans. Heat the oil in a large skillet; add the chopped onion and pepper; sauté for a few minutes.

5. Add the carrots and celery and cook until crisp tender. Add the tomato sauce, basil leaves, and the Parmesan cheese. Mix all ingredients together.
6. Add the mixture to a greased 2-quart casserole dish. Bake uncovered in a 375°F oven until hot and bubbly, about 25 minutes.

3. Western Bean Stew

1 cup pinto beans, dry
3 cups water to soak beans
1/2 teaspoon salt
2–3 drops Tabasco sauce
2 tablespoons shortening
1 medium onion, chopped

1 clove garlic, minced
1 cup canned tomatoes
2 tablespoons flat leaf parsley, chopped
1/2 cup water
1/4 teaspoon dried marjoram
1 teaspoon chili powder

1. Sort and wash beans. Soak overnight.
2. Add salt and Tabasco sauce to beans; bring to a boil; reduce heat; simmer until beans are tender, for approximately 1 hour. Drain beans.
3. While beans are cooking, melt fat in a heavy frying pan; add onions and garlic; cook until the onion is light yellow. Add tomatoes, parsley, 1/2 cup water, and the spices; simmer together for 30 minutes.
4. Add beans to onion mixture; simmer together for an additional 15 minutes.

4. Split Pea Soup

1/2 cup split peas
3 cups water
2 oz. smoked ham
1 medium onion, minced
1 garlic clove, minced

1/2 cup carrot, grated
1 stalk celery, sliced
1/4 teaspoon dry mustard
Salt, ground black pepper, paprika

1. Wash peas and combine with all ingredients except the seasonings in a 3-quart saucepan.
2. Bring mixture to a boil, cover, and lower the heat to simmer the mixture.
3. Cook until the peas are tender.
4. Add seasonings to taste.

5. Lentil Stew

1 tablespoon olive oil
2 medium onions, chopped
1 garlic clove, minced
3 cups water
1 cup dried lentils, sorted, rinsed
1/4 cup flat leaf parsley, chopped
1 bay leaf
1/4 teaspoon coriander
1/2 teaspoon cardamom

1/2 teaspoon ground cumin
1/2 teaspoon salt
1/4 teaspoon ground black pepper
1/8 teaspoon mace
1/8 teaspoon cinnamon
8 oz. white mushrooms cut in half
2 medium potatoes, coarsely chopped
1 can (28 oz.) whole tomatoes, undrained

1. Heat olive oil in a Dutch oven over medium heat.
2. Sauté onion and garlic until the onion is tender.
3. Add the rest of the ingredients except for the potatoes and tomatoes. Bring the mixture up to a boil, lower heat, cover, and cook until the lentils are tender, about 30 minutes. Check occasionally if more water is needed.
4. Add the tomatoes and the potatoes. Continue cooking until the potatoes are tender, about 20 minutes. Add additional water if the mixture looks too thick or starts to stick.

QUESTIONS

1. What foods may legumes be substituted for in the diet?

2. How do legumes compare with other protein-rich foods in nutritive value?

3. What are the basic principles in cooking legumes?

4. What factors influence the cooking time of legumes?

5. a. What is the test for doneness of legumes?

   b. What would happen if the following ingredients were added to legumes before they were fully cooked: tomato sauce; brown sugar?

II. TO LEARN ABOUT AND BECOME ACQUAINTED WITH TOFU IN FOOD PREPARATION

A. TOFU AND TUNA SALAD

1 can (6½ oz.) tuna in water
1½ cups soft tofu
1 cup celery, chopped
1/2 cup onion, chopped

1/2 teaspoon dill seed
Dash ground black pepper
1/2 cup low-fat or light mayonnaise

1. Drain tuna and flake.
2. Thoroughly mix all ingredients. Chill before serving.

B. TOFU SPINACH PIE

1 nine-inch pie shell, unbaked
1 package (10 oz.) frozen chopped spinach, thawed
  and thoroughly drained
1/4 cup olive oil
1 cup onion, chopped

1 pound firm tofu, crumbled
1 teaspoon garlic powder
1 tablespoon lemon juice
1 teaspoon salt

1. Preheat oven to 400°F.
2. Partially bake ("blind bake") the pie shell for 5 minutes. Remove from the oven.
3. Drain the defrosted spinach well. Squeeze thoroughly.
4. Sauté the onions in the olive oil until soft and translucent.
5. Add the spinach to the cooked onions and cook for 2 minutes more.
6. Add crumbled tofu, lemon juice, and garlic powder to the spinach mixture.
7. Spoon mixture into the partially baked pie shell. Bake for about 30 minutes or until crust is golden.

C. CHILI CON TOFU WITH BEANS

1/2 pound frozen tofu, defrosted
1 tablespoon soy sauce
1 tablespoon tomato paste
1/2 tablespoon creamy peanut butter
Dash of onion powder
Dash of garlic powder
2 tablespoons water
2 tablespoons vegetable oil, divided

1 large green pepper, diced
1 large onion, diced
1 garlic clove, minced
1 cup canned tomatoes
$1\frac{1}{4}$ cups cooked pinto beans, with liquid
1 teaspoon salt
$2\frac{1}{2}$ teaspoons chili powder
Dash of cumin

1. Thaw and squeeze the water out of the frozen tofu. Tear into bite-sized pieces.
2. Mix together soy sauce, tomato paste, peanut butter, onion powder, garlic powder, water, and 1/2 tablespoon vegetable oil. Add the tofu to this mixture and mix well until all pieces are evenly coated.
3. In a deep skillet, fry tofu in 1 tablespoon of the vegetable oil over medium heat until brown.
4. In a separate skillet, heat 1/2 tablespoon vegetable oil. Sauté onion, green pepper, and garlic until onions are transparent and green pepper is softened.
5. Add the sautéed vegetables, tomatoes, and beans to the tofu. Add water if needed to cover.
6. Add remaining spices to the tofu and bring to a simmer.
7. Serve with Buttermilk Cornbread (see Laboratory 14).

D. ITALIAN VEGETABLE PIE

6 lasagna noodles, cooked according to package
     directions
1 tablespoon olive oil
1 cup chopped green pepper
1 cup chopped onion
1 cup chopped mushrooms (about 3 oz.)
3 garlic cloves, minced
3 tablespoons tomato paste
1 teaspoon dried Italian seasoning

1 teaspoon fennel seeds
1/4 teaspoon crushed red pepper
1 (12.3 oz.) package firm tofu, drained and
     crumbled
1 (25.5 oz.) bottled marinara sauce
Cooking spray
$1\frac{1}{2}$ cups (6 oz.) preshredded part-skim mozzarella
     cheese
1/4 cup grated Parmesan cheese

1. Preheat oven to 375°F.
2. Cook lasagna noodles according to package directions.
3. While lasagna noodles are cooking, add olive oil to a large skillet. Heat skillet over medium-high heat. Add green pepper, onion, mushrooms, and garlic; sauté for 3 minutes or until vegetables are tender. Stir in tomato paste and next five ingredients; bring to a boil. Reduce heat, and simmer uncovered for 10 minutes.
4. Cut cooked lasagna noodles in half, crosswise. Arrange the cut lasagna noodles spoke like in the bottom and up the sides of a deep 9-inch pie plate coated with vegetable spray.
5. Spread about 3 cups of tofu mixture over lasagna noodles. Fold ends of lasagna noodles over tofu mixture; top with remaining tofu mixture, covering ends of noodles. Sprinkle with the Parmesan and mozzarella cheeses.
6. Bake for 20 minutes; let it stand for 5 minutes before serving.

E. EGGLESS SALAD

1 package (14 oz.) firm tofu, drained
1/2 teaspoon dried dill
1/8 teaspoon cayenne pepper
1/2 teaspoon turmeric
Salt and pepper to taste
3 tablespoons light mayonnaise

1 tablespoon Dijon mustard
1 tablespoon sweet pickle relish
1/8 teaspoon celery salt
2 green onions, minced
1 tablespoon flat leaf parsley, minced
1–2 drops yellow food coloring

1. Drain tofu thoroughly. Place the tofu in a mixing bowl. Mash the tofu with a potato masher or use a wooden spoon.
2. Mix in the remaining ingredients and combine well. Chill slightly, then serve on a bed of lettuce or serve in a sandwich.

F. LOAF CAKE*

2¼ cups cake flour
1 cup sugar
2 teaspoons double-acting baking powder
1 teaspoon salt
¼ cup butter, at room temperature

1/4 cup soft tofu, drained
1 teaspoon vanilla
5 egg yolks
3/4 cup 2% milk

1. Heat oven to 350°F.
2. Grease and line the bottom of a loaf pan, 9 × 5 × 3 inches.
3. Sift together the flour, sugar, baking powder, and salt into a medium-sized mixing bowl.
4. Add butter, tofu, vanilla, egg yolks, and milk. Beat with mixer for 3 minutes, scraping the bowl constantly.
5. Spoon batter into pan. Bake for 60 minutes or until cake tests done with a toothpick.

*This recipe was developed as a class project in Experimental Foods where students used tofu as a partial replacement for fat in a baked product.

G. LEMON CHEESECAKE BARS*

1 cup graham cracker crumbs (about 12 squares)
3 tablespoons butter or margarine, melted
4 oz. cream cheese or low-fat variety, softened
4 oz. silken tofu
1/2 cup granulated sugar

1/4 cup soymilk
1 teaspoon grated lemon rind
1 teaspoon vanilla extract
3 eggs

1. Heat oven to 325°F. Mix cracker crumbs and butter thoroughly. Press evenly in the bottom of an ungreased rectangular baking dish, 11 × 7 × 1½ inches. Bake for about 10 minutes.
2. Beat cream cheese until creamy; add the silken tofu. Mix until the cheese and tofu are incorporated.
3. Beat in sugar, milk, lemon rind, and vanilla extract.
4. Beat in eggs, one at a time. Spread mixture over crust.
5. Bake about 30 minutes or until center is set. Cool thoroughly. Cover and refrigerate for at least 2 hours. Cut into 2¼ × 1¼ inch bars. Refrigerate any leftovers.

*This recipe was developed as a class project in Experimental Foods where students used tofu as a partial replacement for fat in a baked product.

QUESTIONS

1. Briefly describe the source and preparation of tofu.

2. What happens to tofu when it is frozen?

3. What are the commercial forms of available tofu, and how are they used in food preparation?

4. What are the nutritional advances made with regard to soybeans and tofu?

LABORATORY 17

Sugar Crystallization

LABORATORY 17
Sugar Crystallization

Candy making is an art which requires accuracy in measuring and cooking. Sugar plays an important role in developing the texture and consistency of candy. Cooking the sugar solution to the proper point will insure a quality product. This laboratory exercise will introduce the student to the different types of candy and the preparation techniques that are used to make the particular varieties.

VOCABULARY

amorphous candy	fructose	saturated solution
caramelizing sugar	invert sugar	seeding
corn syrup	interfering agent	sucrose
crystalline candy	monosaccharide	supersaturated solution
disaccharide	nuclei sites	

OBJECTIVES

1. To better understand the principles involved in the successful preparation of various types of candies.
2. To observe and understand the difference between crystalline candy and amorphous candy.
3. To understand the manipulation of heating, cooling, and beating, and the effect they will have on crystalline candy.

PRINCIPLES

1. An invert sugar is a mixture of equal amounts of glucose and fructose resulting from the hydrolysis of sugar.
2. Hydrolysis of sugar occurs through heating, acid, or through the use of an enzyme, invertase.
3. To make most candies, sucrose solutions (saturated solution) are boiled until enough water is evaporated to produce a sugar concentration (supersaturated solution) that yields a candy of the desired consistency.
4. Crystalline candies contain sucrose crystals. These crystals should be so small in the finished product that they go undetected by the tongue. Examples of crystalline candies include fondant, fudge, penuche, divinity, marshmallows, creams, and nougats.
5. In order for the crystals, sucrose must be completely dissolved and must be cooked to the correct concentration.
6. The steps necessary to obtain an acceptable crystalline candy are:
   a. *Heating*: boil the sucrose solution (saturated solution) in an open pot to evaporate the excess water.
   b. *Cooling*: a supersaturated solution is necessary to produce small crystals in the candy. As the candy cools, it becomes more viscous and more unstable. However, this viscous texture favors formation of small nuclei when crystallization occurs.
   c. *Agitation*: once a supersaturated solution is obtained, the viscous syrup must be rapidly beaten to produce numerous small crystals. Beat until dull. Overbeating will cause large sugar crystals.
7. Interfering agents in the sugar solution inhibit the formation of large crystals. Interfering agents include corn syrup, butter, milk, cream, chocolate, cocoa, gelatin, and egg white. They have a different structure than sucrose and physically interfere with sucrose crystals growing on each other.
8. Noncrystalline candies are also known as amorphous candies because they do not contain crystals. Such candies either contain large amounts of interfering agents or have been cooked to high end-point temperatures, evaporating all of the water.

ATTENTION

Before beginning to make candy, check the boiling point of the thermometer you are using. Adjust the temperatures in the recipes according to your results. For example, if water boiled at 210°F (99°C), lower temperature given in the recipes by 2°F (1°C).

I. TO SUCCESSFULLY PREPARE CRYSTALLINE CANDY BY LEARNING THE PRINCIPLES OF PROPER MANIPULATION

A. CHOCOLATE FUDGE

2 cups granulated sugar
2/3 cup whole milk
2 tablespoons light corn syrup
1/4 teaspoon salt

2 oz. unsweetened chocolate, chopped
2 tablespoons butter or margarine
1 teaspoon vanilla extract
1/2 cup chopped walnuts, if desired, coarsely chopped

1. Butter an 8 × 8 × 2 inch pan. Mix together sugar, milk, corn syrup, and salt in a 2-quart saucepan. Be careful when stirring not to get the sugar on the sidewalls of the pan. Add the chopped chocolate, and start heating the mixture, stir while it is heating until the chocolate melts. When the sugar mixture starts boiling, cover the pan and allow the mixture to boil covered for 2 minutes.
2. Uncover the saucepan, and attach a candy thermometer allowing the bulb to be immersed in the boiling mixture. Do not allow the bulb to touch the bottom of the pan. Cook the mixture until it reaches 234°F (112°C) on the candy thermometer or until a small amount of the mixture forms a softball when dropped into some ice water.
3. Remove from heat. Add butter. **Do not mix**.
4. Cool the mixture, undisturbed, until it reaches 120°F (49°C). Add vanilla.
5. Beat the mixture vigorously with a wooden spoon until it loses it shine. Quickly stir in walnuts.
6. Immediately transfer the candy to prepared pan. Cool until firm. Cut into 1-inch squares.

B. PEANUT BUTTER FUDGE

2 cups granulated sugar
2 cups brown sugar
1 teaspoon salt
1/2 cup natural cocoa

$1\frac{1}{3}$ cups milk
1/4 cup butter
3/4 cup creamy peanut butter
2 teaspoons vanilla extract

1. In a Dutch oven sift together the cocoa and the granulated sugar. Add the brown sugar, salt, and milk.
2. Heat until the sugar is dissolved. Add the butter and boil the mixture until a temperature of 238°F (115°C) is reached on a candy thermometer. Also add a teaspoon of the mixture to ice water and test if it will form a softball.
3. Pour mixture into a large mixing bowl. Add the vanilla and peanut butter. With a paddle attachment start mixing at medium speed until the fudge mixture lightens and looses its shine.
4. Pour it into a greased 9 × 9 × 2 inch pan. Let it cool completely and cut into $1\frac{1}{2}$ inch squares.

C. PENUCHE

1 cup granulated sugar
1 cup light brown sugar, packed
2/3 cup milk
2 tablespoons light corn syrup

1/4 teaspoon salt
2 tablespoons butter or margarine
1 teaspoon vanilla extract
1/2 cup chopped walnuts, if desired

1. Butter an 8 × 8 × 2 inch pan; set aside.
2. Combine in a 2-quart saucepan the granulated sugar, brown sugar, milk, corn syrup, and salt. Stir to combine, being careful not to get the mixture on the sidewalls of the pan.
3. Start heating the mixture. Place a cover on the pan and allow the mixture to boil covered for 2 minutes to wash down the sidewalls of the pan.
4. Remove the cover and place a candy thermometer into the pan. Do not allow the bulb to touch the bottom of the pan. Cook the mixture to 234°F (112°C) or until a small amount of the mixture forms a softball when dropped into ice water.
5. Remove from heat. Add butter. Cool mixture to 120°F (49°C) without stirring.
6. Add vanilla. Beat vigorously and continuously for 5–10 minutes until the candy is thick and no longer glossy.
7. Quickly stir in nuts, if desired. Spread in pan. Cool until firm. Cut into 1-inch cubes.

D. DIVINITY

1 cup sugar
2 tablespoons light corn syrup
1/4 cup boiling water

1 egg white
Dash of salt
1/2 teaspoon vanilla extract

1. Mix first three ingredients in a 1½-quart saucepan and stir until sugar is dissolved, or heat for 2–3 minutes in a covered pan.
2. Remove lid and boil rapidly to 252°F (122°C).
3. Add salt to egg white and beat until stiff peaks but shiny.
4. In a steady stream, pour the hot syrup into the beaten whites, beating continuously.
5. Beat until thick and stiff enough to hold its shape and until candy begins to lose its gloss.
6. Add vanilla and drop onto wax paper.

Table 17.1 EVALUATION OF CRYSTALLINE CANDY				
Crystalline Candy	Firmness	Crystal Size	Smoothness	Flavor
Chocolate fudge				
Penuche				
Divinity				

QUESTIONS

1. What is invert sugar and how can it be formed in the candy-making process?

2. What is the difference between a saturated and a supersaturated sugar solution?

3. Why must a supersaturated solution be obtained when making crystalline candies?

4. How must a supersaturated solution be handled in order to insure a quality candy product?

5. Why is it necessary not to get sugar on the sidewalls of the cooking vessel when making candy?

6. Why is it necessary not to jar the cooling candy mixture?

7. What are interfering agents in candy making? Identify the agents used in the candy recipes in this manual.

II. TO OBSERVE AND SUCCESSFULLY PREPARE AMORPHOUS CANDY

A. CARAMELS

1 cup granulated sugar
1 cup dark corn syrup
1 cup light cream, divided

1/2 cup butter
1 teaspoon vanilla extract

1. Lightly butter the sides and bottom of an 8 × 8 × 2 inch pan.
2. In a 2-quart saucepan blend together the granulated sugar, corn syrup, 1/2 cup light cream, and butter. Bring to a boil stirring constantly.
3. Cook over moderate heat, stirring constantly to 240°F (116°C).
4. Remove from heat and gradually add the balance of the light cream (1/2 cup).
5. Return the mixture to the heat and cook to 244–246°F (118–119°C). Stir in quickly the vanilla extract.
6. Pour the mixture at once, without stirring, into the buttered pan. Allow the caramels to set until cool. Loosen from sides of pan with a flat edge knife. Invert the candy onto a wooden board.
7. Cut with lightly buttered knife; wrap in waxed paper.

B. PEANUT BRITTLE

3 tablespoons butter or margarine, divided
1½ cups granulated sugar
1/2 light corn syrup
1/2 cup water

1/2 teaspoon baking soda
1½ cups roasted peanuts, coarsely chopped
1/2 teaspoon vanilla extract

1. Use 1 tablespoon butter to lightly grease the surface of three large baking sheets without sides.
2. Using a 2-quart saucepan, combine the sugar, water, and corn syrup. Place the thermometer in position.
3. Heat mixture rapidly to 280°F (138°C). Stir to keep mixture from scorching. It may be desirable to wipe undissolved sugar crystals from the side of the pan with damp cheesecloth wrapped around a fork.
4. When the syrup reaches 280°F (138°C), add the peanuts and the balance of the butter. Stir the mixture continuously and heat to 306°F (152°C). Remove the pan immediately from the heat. **Remember to make temperature adjustments for original thermometer reading for boiling water; if water boiled at 210°F (99°C), cook to 304°F (151°C).**
5. Add the baking soda and the vanilla; stir these ingredients in as quickly as possible. **Do not over stir or the foam structure will be lost.**
6. Pour approximately one-third of the final mixture onto each of the three baking sheets. Pour into as thin a layer as possible, but do not try to spread the mixture with the spatula.
   After the edges of the candy have cooled slightly—about 2 minutes—start to gently pull and stretch the candy into a relatively thin sheet. Try to keep the nuts fairly evenly distributed during stretching. Continue to stretch the candy until the center of the mass has also been stretched.
7. When completely cooled, break into pieces.

C. TOFFEE

1 cup granulated sugar
1/2 cup butter
1/4 cup water

1 tablespoon light corn syrup
1/2 cup blanched almonds, toasted, chopped
1/8 pound milk chocolate

1. Mix sugar, butter, water, and corn syrup in a 2-quart saucepan and stir until sugar is dissolved.
2. Cook on medium heat to hard-crack stage (300°F or 149°C), stirring to prevent scorching.
3. Stir in almonds and pour into a buttered 8 × 8 × 2 inch pan.
4. When candy is cool, spread with chocolate which has been melted in a double boiler over warm water (115°F or 46°C) water.
5. When cold, break the coated toffee into pieces.

Table 17.2 EVALUATION OF AMORPHOUS CANDY				
Amorphous Candy	Firmness	Crystal Size	Smoothness	Flavor
Caramels				
Peanut brittle				
Toffee				

III. MISCELLANEOUS CANDY

A. CHOCOLATE TRUFFLES

4 oz. bittersweet chocolate, grated
4 oz. semisweet chocolate, grated
1/2 cup heavy cream
1½ tablespoons Kaluha*

1 tablespoon prepared coffee
1/2 teaspoon vanilla extract
Confectioners' sugar
Cocoa powder

1. Heat heavy cream in a small saucepan until it boils.
2. Immediately pour cream through a sieve into a bowl with the grated chocolate. With a wire whisk slowly stir the cream and chocolates together until the chocolate is completely melted (if the chocolate does not melt, place the bowl over a pan of simmering water).
3. Whisk in flavors (Kaluha, vanilla, and coffee).
4. Place mixture in a glass $8 \times 8 \times 2$ inch pan. This will speed up the cooling process. Cover and chill for 30–45 minutes until pliable but firm.
5. With 2 teaspoons or 1¼ inch ice cream scoop or melon baller make dollops of the chocolate mixture and place on a baking sheet lined with parchment paper. Refrigerate for 15 minutes.
6. Roll in confectioners' sugar or cocoa.

*Variation: These candies can be made with a variety of flavors. In place of Kaluha, Cointreau could be used, as well as raspberry, almond, or hazelnut liquors.

B. MICROWAVE NUTTY CARAMEL CORN

8 cups popped unflavored popped corn
2 cups light brown sugar, packed
1/2 cup light corn syrup
1 teaspoon salt

8 tablespoons butter
2 teaspoons vanilla extract
1 teaspoon baking soda
1½ cups roasted unsalted peanuts

1. Prepare popcorn according to package directions; set aside.
2. In a 1-quart glass measuring cup combine the brown sugar, corn syrup, salt, butter, and vanilla. Microwave on HIGH for 1 minute. Stir; microwave on HIGH for another 2 minutes. Add the baking soda and stir well (mixture will foam).
3. Add the prepared popcorn to a large brown paper bag. Combine the peanuts with the microwave caramel sauce; pour it over the popcorn, and with a spoon with a long handle stir to coat.
4. Turn the bag under to close; microwave on HIGH for 1 minute.
5. Transfer the caramel corn mixture onto a baking sheet lined with waxed paper or parchment paper to cool. Store mixture in an airtight container.

QUESTIONS

1. What are the interfering agents in caramels, and how do they function?

2. Why was baking soda added to the peanut brittle and how does it affect the texture of the candy?

3. What is the main difference between amorphous and crystalline candies?

4. What determines the consistency of fudge? Caramels?

5. What process is responsible for the production of color in:

   a. peanut brittle?

   b. caramels?

LABORATORY 18

Ice Crystallization
(Frozen Desserts)

LABORATORY 18
Ice Crystallization (Frozen Desserts)

On a hot summer day, people always look for something cold and refreshing. Ice cream has been a favorite selection for such an occasion. Body and texture are terms used when rating ice cream. Sugar plays a supporting role, along with the other ingredients in the recipe to insure proper body and texture. The student will learn in this laboratory exercise how the selection and manipulation of the ingredients in the ice cream recipe will enhance body and texture.

VOCABULARY

body	French vanilla ice cream	stabilizer
brine	overrun	still frozen
emulsifier	sherbet	texture

OBJECTIVES

1. To understand the functional role of ingredients in an ice cream formulation.
2. To understand how the rate of freezing will affect the body of the ice cream.
3. To differentiate between body and texture when describing quality in a frozen dessert.

PRINCIPLES

1. The ingredients used in a frozen dessert influence body, texture, consistency, and flavor.
2. Sugar absorbs moisture in the recipe; therefore, it produces a smooth texture. Also, the amount of sugar lowers the freezing point of the mixture, thereby affecting the body of the frozen dessert. Sugar also contributes flavor.
3. Cream and milk provide fat; the more fat in the mixture the smoother the texture; the fat interferes with crystal growth or formation. The fat also affects the body of the ice cream.
4. Emulsifiers, such as mono- and diglycerides, are used to keep the fat dispersed. Stabilizers, such as guar gum and locust bean gum, help absorb water and increase the viscosity of the mix. These ingredients aid in maintaining the body and texture of the ice cream.
5. Overrun is the increase in volume of a frozen mixture when it is frozen. While part of the increase in volume results in the expansion of water during freezing, most of it is due to the incorporation of air in the mixture during freezing.
6. The time when the mixture freezes will affect the overrun. If the mixture freezes too fast not enough air will be incorporated; therefore, the body will be heavy. If the freezing takes a long time too much air will be incorporated; therefore, the body will be light.
7. Still-frozen desserts are not agitated during their freezing, but are whipped after they are partially frozen. They do not have the same texture and body as the agitated frozen dessert, and ice crystals grow faster in these types.

I. PREPARATION OF FREEZER AND FREEZER (MIXTURES)

1. See that all parts of the freezer fit and are in working order.
2. Scald freezer can and dasher.
3. Pour mixture (should be made ahead and chilled thoroughly) to be frozen into freezer can. Do not fill more than 2/3–3/4 full.
4. Adjust lid and dasher and place in freezer container (the container should be chilled in a refrigerator freezer prior to making the ice cream).
5. Fill freezer container with 1/3 crushed ice; add 1 cup of rock salt around the top of the ice. Continue this procedure alternating layers of ice and rock salt (should be three layers of each). **This step is important since this will affect the freezing rate of the ice cream mixture.**

Freezing Process

6. Follow manufacturer's guide for freezing and operating the machine that you have.

179

Packing the Dessert

7. Remove the dasher and pack down the frozen dessert into another bowl.
8. Cover the surface of the ice cream with a piece of waxed paper.
9. Cover bowl entirely; place in freezer and allow the ice cream to "ripen" to develop flavor.

Observations

1. Determine the temperature of the mix and of the brine and record it in Table 18.1.
2. Determine the **swell** or **overrun** as follows:
   a. Measure the depth of the can (A).
   b. Measure from top of can to top of mix before freezing (B); subtract to get depth of mix (A − B = C).
   c. As soon as the mixture is frozen, remove dasher and measure to top of the frozen mix (D).
   d. Subtract original depth of mix to get the amount of swell (C − D).
   e. Divide the amount of swell by original depth and multiply quotient by 100; this gives percentage (%) of swell or overrun.
   f. Formula:

$$\% \text{ swell or overrun } = \frac{C - D}{C} \times 100$$

Table 18.1 EVALUATION OF QUALITY ATTRIBUTES OF FROZEN DESSERTS						
Frozen Dessert	T° of Brine	T° of Mix	% Swell	Body	Texture	Flavor
French vanilla ice cream						
Old-fashioned lemon ice cream						
Chocolate ice cream						
Strawberry ice cream						
Orange sherbet						

A. FRENCH VANILLA ICE CREAM

1/2 teaspoon unflavored gelatin plus 1 tablespoon water
2 large egg yolks
1/2 cup granulated sugar
1 cup whole milk

1/4 teaspoon salt
1 tablespoon vanilla extract
1 vanilla bean, split and scrapped of its seeds (optional)
2 cups heavy cream

1. Mix together unflavored gelatin and water in a custard cup and set aside.
2. Whisk together the egg yolks, sugar, milk, and salt in the top of a double boiler.
3. Cook mixture over simmering water until it coats a spoon. Stir in gelatin and stir until gelatin dissolves. Add vanilla and vanilla bean seeds (if desired).
4. Place custard into a bowl and chill the mixture thoroughly. **The custard could be made a day in advance. The colder it is the better for freezing into ice cream.**
5. Stir the heavy cream into the chilled custard mixture.
6. Pour it into the chilled freezer container; put dasher in place.

7. Freeze mixture according to manufacturer's instructions.
8. Allow to ripen for several hours before serving.

B. OLD-FASHIONED LEMON ICE CREAM

1/4 teaspoon salt
1½ cups granulated sugar
2 tablespoons + 2 teaspoons all-purpose flour
2 cups whole milk
2 large egg yolks, beaten

1 cup half and half
1 cup heavy cream
2 teaspoons vanilla extract
1/4 teaspoon grated lemon rind
1/4 cup + 2 tablespoons freshly squeezed lemon juice

1. Combine salt, sugar, and flour in a 2-quart saucepan.
2. Gradually add milk; stir until dry ingredients are blended and dissolved.
3. Cook over medium heat, stirring constantly, until thickened and the mixture comes to a boil.
4. Gradually stir in approximately 1/4 of the hot mixture into the 2 beaten egg yolks. Add this mixture to the remaining hot mixture in the saucepan. Stir thoroughly until blended. Cook for 1 minute; remove from heat and allow mixture to cool. Cover and cool for 2–24 hours (for faster results place mixture over ice to cool rapidly).
5. Add half and half, heavy cream, vanilla, and grated lemon rind to the chilled mixture.
6. Pour in lemon juice and beat well.
7. Pour mixture into freezer container. Freeze according to manufacturer's directions.

C. CHOCOLATE ICE CREAM

1/2 teaspoon unflavored gelatin
2 cups whole milk, divided
2 large egg yolks
1 cup granulated sugar
2 cups heavy cream

2 teaspoons vanilla extract
3 oz. semisweet chocolate, grated
2 oz. unsweetened chocolate, grated
1/3 cup chocolate chips, optional

1. Dissolve gelatin in a 1/2 cup milk; set aside.
2. Scald balance (1½ cups) of the milk in the top of a double boiler. Add the gelatin.
3. Beat the egg yolks with the sugar. Gradually add some of the scalded milk to temper the yolks. Mix thoroughly.
4. Transfer the mixture back to the remaining milk in the double boiler.
5. Heat until the mixture coats a spoon.
6. At this point cool the mixture thoroughly over ice to quicken the cooling process.
7. Microwave the grated chocolates on HIGH for 20 seconds. Remove from the microwave and stir. Return to the microwave and heat on HIGH for 25 seconds. Remove and stir. Return to microwave and heat again if necessary for another 10–20 seconds until thoroughly melted.
8. Add the melted chocolate, heavy cream, and vanilla to the chilled custard.
9. Add the custard to the freezer container. Freeze according to manufacturer's directions.
10. After 15 minutes freezing, add the chocolate chips.
11. Continue freezing. Place the ice cream in a container and harden before serving.

D. STRAWBERRY ICE CREAM

1. Follow ingredients and directions for French Vanilla Ice Cream, except:
   a. omit vanilla bean seeds.
   b. decrease vanilla extract to 1 teaspoon.
   c. stir in one package (16 oz.) frozen strawberry halves, thawed, into custard mixture after adding the vanilla extract.
   d. stir in 2–3 drops of red color, if desired.

E. ORANGE SHERBET

1/2 cup + 2 tablespoons water
1/2 teaspoon unflavored gelatin
1/2 cup granulated sugar
1/2 cup whole milk

1/2 cup freshly squeezed orange juice
1 tablespoon freshly squeezed lemon juice
1/2 teaspoon orange rind, grated
Dash of salt

1.   Hydrate unflavored gelatin in 2 tablespoons water.
2.   Heat 1/2 cup water with sugar; add hydrated gelatin. Stir over heat to dissolve ingredients.
3.   Add the remaining ingredients.
4.   Chill the mixture thoroughly before freezing.
5.   Freeze according to manufacturer's directions.

NOTE: This recipe needs to be doubled in order to fit into the normal ice cream freezer. If using the smaller hand-cranked models that are placed directly into the freezer, this recipe "as is" would be perfect.

QUESTIONS

1.   Distinguish between ice cream, low-fat ice cream, fat reduced ice cream, sherbet, sorbet, and ices.

2.   Why is rock salt combined with ice to freeze dessert mixes?

3.   Briefly discuss the functions of sugar and fat in frozen dessert mixes.

4.   What ingredients are used in low-fat and fat-free ice cream to create body and texture?

5.   a.   What causes swell or overrun in ice cream?

     b.   Why is it important and how does it affect the eating quality of the product?

6.   What was the role of unflavored gelatin in the recipes used in this unit?

7.   a.   What would cause sandiness in ice cream?

     b.   How is this different from iciness?

8.   What do frozen desserts and crystalline candies have in common?

II. TO USE STILL-FREEZING IN THE MAKING OF FROZEN DESSERTS

A. ITALIAN LEMON ICE

Juice of 6 lemons
$2\frac{1}{2}$ cups water
1 cup granulated sugar

Grated rind of 1 lemon
1 egg white

1. Squeeze lemons and discard seed.
2. Combine water and sugar in saucepan. Bring to a boil and boil gently for 5 minutes; cool.
3. Stir in lemon juice and grated rind. Using a whisk, beat in egg white.
4. Pour mixture into a wide shallow pan (this will allow the mixture to cool down quickly) and place in freezer.
5. When lemon ice is semifrozen, remove from freezer. Transfer the mixture to a bowl and beat for 1 minute **(this procedure develops the texture of the ice)**.
6. Pour back the Italian ice into a container and freeze.

B. CHOCOLATE MOUSSE

$1\frac{1}{2}$ cups milk
3 oz. (3 squares) semisweet chocolate, grated
1/4 cup granulated sugar
3 large eggs, separated

1 teaspoon vanilla extract
1/2 teaspoon cream of tartar
3/4 cup heavy cream, whipped

1. In a saucepan over low heat stir and scald the milk, sugar, and chocolate. Heat until chocolate melts.
2. Remove from heat. Add a small amount of the chocolate mixture to the beaten egg yolks. Add the yolks back to the saucepan and heat the custard over low heat; stirring constantly until it thickens and custard coats a spoon. Strain the custard. Cool in a bowl over ice. Add vanilla extract.
3. Beat egg whites and cream of tartar in a $2\frac{1}{2}$-quart bowl until stiff, but shiny peaks.
4. Stir about 1/4 of the egg white foam into the chocolate mixture. Fold mixture into remaining egg white foam.
5. Fold whipped heavy cream into the chocolate egg white foam. Spoon into dessert dishes.
6. Refrigerate at least 2 hours before serving (see Table 18.2).

Table 18.2 EVALUATION OF STILL-FROZEN DESSERTS			
Still Frozen Dessert	Body	Texture	Flavor
Italian lemon ice			
Chocolate mousse			

QUESTIONS

1. What is a mousse?

2. How is air incorporated into still-frozen desserts?

3. How is crystal growth prevented in still-frozen desserts?

LABORATORY 19

Beverages: Coffee, Tea, and Cocoa

Coffee, tea, and cocoa are prepared and enjoyed worldwide. Aroma and body are key quality factors that are usually associated with these products. Careful preparation is important in order to enjoy these products to their fullest. This laboratory exercise will introduce the student to the selection of various coffees, teas, and cocoa/chocolate products and the proper preparation of each.

VOCABULARY

black tea	Dutch-processed cocoa	oolong tea
Coffea arabica	freeze-dried	percolation
caffeine	green tea	polyphenol
conching	infusion	tannin
decaffeinated	natural-processed cocoa	white tea
drip method		

OBJECTIVES

1. To learn differences between the various coffees, teas, and cocoas available.
2. To emphasize the main factors in the preparation of tea, coffee, and hot chocolate.

PRINCIPLES

1. Coffee beans are produced by the *Coffea* plant. The two most commonly grown species of the coffee plant are *Coffea canephora* and *C. arabica*.
2. Arabica coffee (from *C. arabica*) is considered more suitable for drinking than robusta coffee (from *C. canephora*); robusta tends to be bitter and have less flavor than arabica. For this reason, about three-fourths of coffee worldwide is *C. arabica*.
3. Once ripe, coffee beans are dried. The seeds are roasted which influences the taste of the beverage produced by changing the coffee bean both physically and chemically.
4. Depending on the color of the roasted beans, they are labeled from light, medium, and dark (French or Italian roast). Darker roasts are generally smoother, because they have less fiber content and a more sugary flavor. Lighter roasts have more caffeine, resulting in a slight bitterness, and a stronger flavor from aromatic oils and acids which are destroyed by stronger roasting.
5. The type of grind is often named after the brewing method for which it is generally used. Turkish grind is the finest grind, while coffee percolator or French press is the coarsest grind. The most common grinds are between the extremes; a medium grind is used in most common home coffee brewing machines.
6. Coffee is best brewed between 185 and 203°F (85–95°C). Polyphenols are more soluble at boiling and give a bitter product.
7. A clean coffee pot is essential in making a good coffee beverage.
8. Tea is a beverage made by steeping processed leaves, buds, or twigs of the tea bush (*Camellia sinensis*).
9. The leaves are processed by oxidation, heating, drying, and the addition of other herbs, flowers, spices, and fruits. The term "herbal tea" usually refers to the infusions of fruit or of herbs (such as rosehip, chamomile, or *jiaogulan*) that contain no *C. sinensis*.
10. Tea is traditionally based on producing technique:
    a. *White tea*: young leaves (new growth buds) that have undergone no oxidation; the buds may be shielded from sunlight to prevent formation of chlorophyll.
    b. *Green tea*: the oxidation process is stopped after a minimal amount of oxidation by application of heat, either with steam, or dry cooking in hot pans.
    c. *Oolong*: oxidation is stopped somewhere between the standards for green tea and black tea. The oxidation process takes between 2 and 3 days.
    d. *Black tea*: the leaves are allowed to completely oxidize.

11. Teas that have little or no oxidation period, such as white or green tea, are best brewed at lower temperatures around 176°F (80°C), while teas with longer oxidation periods should be brewed at higher temperatures around 212°F (100°C).
12. Cocoa and chocolate are made by grinding the seeds of the cocoa tree. To decrease the bitter taste, the seeds are first fermented and dried. The nibs are removed and roasted to develop the flavor further.
13. Cocoas may be divided into two main classes: natural-processed and Dutch-processed.
14. Chocolate has more fat than cocoa, and when substituting, 1 oz. unsweetened chocolate = 3 tablespoons cocoa + 1 tablespoon fat.

I. TO SHOW SOME FACTORS WHICH AFFECT THE QUALITY OF COFFEE BEVERAGES

A. ELECTRIC PERCOLATED COFFEE

2 tablespoons regular grind coffee per cup of water

1. Put cold water in pot. Make a pot at least 2/3 full.
2. Put coffee in strainer, and adjust the strainer in the pot.
3. Cover and plug in the pot.
4. Pot will go through its cycle and percolate the coffee.
5. After percolator stops, allow the coffee to ripen for 5 minutes before tasting.

B. AUTOMATIC DRIP COFFEE

1 tablespoon automatic drip-grind coffee per cup of water

1. Fill pot with desired amount of water.
2. Pour water into the top part of the coffeemaker.
3. Remove filter basket; insert filter and desired amount of coffee.
4. Place filter basket into position.
5. Plug in coffeemaker and turn on.
6. After coffeemaker has gone through its cycle, allow coffee to ripen for 5 minutes before tasting.
7. Serve within 30 minutes.

C. PREPARE DRIP COFFEE AND PERCOLATED COFFEE USING 2 TABLESPOONS OF COFFEE PER CUP OF WATER

Rank and compare coffees made by these two methods for the characteristics listed in Table 19.1.

Table 19.1 EVALUATION OF PERCOLATED AND DRIP COFFEE				
Method	Brownness	Clarity	Stimulating Aroma	Fresh, Mellow Taste*
Electric percolated				
Automatic drip				

*Absence of bitter or flat taste.

QUESTIONS

1. What were the differences between the appearance of the ground coffees used for each method?

2. Which coffee had the most bitter flavor? Why?

D. INSTANT COFFEE COMPARED WITH REGULAR COFFEE

Prepare a small pot of medium strength coffee following the directions on the label.

1. Regular instant
2. Decaffeinated instant

Compare instant coffee with that from the automatic drip grind. Rank for the characteristics listed in Table 19.2.

Table 19.2 EVALUATING INSTANT COFFEE VERSUS DRIP COFFEE			
Coffee	Stimulating Aroma	Strength	Fresh, Mellow Taste*
Automatic drip grind			
Regular instant			
Decaffeinated instant			

*Absence of bitter or flat taste.

QUESTIONS

1. How does the quality of freshly prepared coffee compare with that of instant coffee?

2. What effect, if any, does removal of the caffeine have upon the beverage?

3. How is instant coffee manufactured?

II. TO SHOW SOME FACTORS WHICH AFFECT THE QUALITY OF TEA

A. STRENGTH OF TEA

Prepare small pots of tea using the following amounts of tea per cup of boiling water:

1. 1/4 teaspoon
2. 1/2 teaspoon
3. 1 teaspoon
4. 2 teaspoons
5. 1 tea bag

Pour boiling water over tea and allow the tea to steep for 2–4 minutes. Pour half of the tea into cups for tasting hot and the other half over crushed ice in glasses for iced tea. Rank the teas for optimum strength in Table 19.3.

Table 19.3 EVALUATING THE EFFECT OF TEMPERATURE AND AMOUNT ON TEA STRENGTH		
Amount of Tea	Hot	Cold
1/4 teaspoon		
1/2 teaspoon		
1 teaspoon		
2 teaspoons		
1 tea bag		

QUESTIONS

1. Account for the effects of the serving temperature on the amount of tea needed for optimum strength.

2. What amount of tea (in teaspoons) would be equivalent to one tea bag?

B. KIND OF TEA

Prepare small pots of tea, using 1 teaspoon of tea per cup of water as follows:

1. Green tea
2. Oolong tea
3. Black tea
4. White tea

Rank green, oolong, black, and white tea for the characteristics listed in Table 19.4.

Table 19.4 EVALUATION OF DIFFERENT TEA VARIETIES				
Tea	Color	Aroma	Astringency	Briskness
White				
Green				
Oolong				
Black				

1. What is the main difference between white, green, oolong, and black tea?

III. TO EVALUATE COCOA AND CHOCOLATE

A. CHOCOLATE

1/2 oz. unsweetened chocolate	2 cups whole milk
2 tablespoons granulated sugar	1/4 teaspoon vanilla extract
1/2 cup water	

1. Combine chocolate, sugar, and water in a small saucepan. Heat to boiling with constant stirring. Continue boiling while stirring until a smooth paste forms. **CAUTION: Do not scorch mixture as it reaches consistency of a paste.**
2. Add milk; heat to 200°F (94°C); add vanilla; serve.

B. COCOA

1 tablespoon natural cocoa	1 cup whole milk
1 tablespoon granulated sugar	1/4 teaspoon vanilla extract
1/4 cup water	

1. Mix together cocoa and sugar in a small saucepan; add water gradually; blend.
2. Heat to boiling with constant stirring. Continue boiling and stirring until a smooth paste forms. **Do not allow the mixture to scorch as it reaches the consistency of a paste.**
3. Add milk; heat to 200°F (94°C); stir in vanilla; serve (see Table 19.5).

Table 19.5 EVALUATION OF HOT CHOCOLATE			
Type	Color	Flavor	Body
Chocolate			
Cocoa			

QUESTIONS

1. What are the desired characteristics for a hot chocolate product?

2. Why was it important to keep the temperature below 200°F during preparation?

GENERAL QUESTIONS

1. What constituents in coffee contribute to the following:

    a. aroma?

    b. flavor?

    c. stimulating quality?

    d. color?

2. a. What type of deterioration occurs in roasted coffee?

    b. Under what conditions is deterioration retarded?

3. What constituents in tea contribute to the following:

    a. aroma?

    b. appearance?

    c. flavor?

    d. stimulating quality?

4. What food constituents are found in cocoa?

5. What is the difference in color between natural cocoa and Dutch-processed (European) cocoa and how is it achieved?

6. What is the weight of the sections into which an 8 oz. package of unsweetened chocolate is divided?

7. A recipe calls for 2 oz. of unsweetened chocolate. You are out of chocolate, but have cocoa. Can a substitution be made? If so, give directions.

8. How should chocolate be stored? What is meant by "chocolate bloom?"

9. Why should cocoa powder and chocolate be cooked in making beverages?

LABORATORY 20

Sensory Evaluation of Food

When the quality of food is judged or evaluated by the senses (flavor, aroma, color, and texture), it is said to be sensory evaluation. Flavor of food is affected by temperature, color, and texture. This laboratory exercise will illustrate to the student how these different perceptions affect the identification and acceptability of a particular food item.

VOCABULARY

aftertaste

main taste sensations

papilla

taste receptors

I. TO STUDY THE EFFECT OF TEMPERATURE ON FLAVOR INTENSITY

The tongue is the main receptor which is effective in determining flavor of food. Temperature will also affect how flavor is perceived by the tongue.

Directions: Scoop vanilla ice cream into three separate cups. Taste the samples using the following temperatures:

-15°C (5°F): Sample #1
5°C (41°F): Sample #2
24°C (75°F): Sample #3

Rank the three samples by placing a check in the appropriate box in Table 20.1. Remember to rinse your mouth between sampling.

Table 20.1 EVALUATION OF SWEETNESS INTENSITY			
Sweetness Intensity	Sample #1	Sample #2	Sample #3
Most sweet			
Moderately sweet			
Least sweet			

QUESTIONS

1. What effect did temperature have on the perceived sweetness of the ice cream?

2. Where is the sweetness sensation found on the tongue?

3. What other sensory characteristic of ice cream is affected by sugar?

II. TO LEARN HOW COLOR AFFECTS FLAVOR

Color plays a central role in the evaluation of food. It not only influences the senses of taste and smell, but also the acceptability of a food product.

Directions: Have the judge sit in a sensory booth and either be blindfolded or obscure the light with a piece of red cellophane paper. Present to the judge five fruit drinks that are different in color. Have the juice identified by flavor only since color will be obscured. Rinse between each sample. In Table 20.2, the judge will identify and record the juice in the order of presentation, and the reason for its selection.

Table 20.2 IDENTIFYING VARIOUS JUICE DRINKS BY FLAVOR	
Juice	Reason for Identifying
1.	
2.	
3.	
4.	
5.	

QUESTIONS

1. What influence does color or appearance have on

   a. taste perception?

   b. product acceptability?

III. TO DETERMINE HOW TEXTURE AFFECTS FOOD IDENTIFICATION

A food's textural attribute can also contribute to its identity and quality. Although a person's awareness of texture is not as apparent as their awareness of other food properties, when vast changes in texture are made, identification of foods can become difficult.

Directions: Present to the judges various foods that have been pureed (e.g., carrots, peas, prunes, etc.) or are naturally soft (e.g., cream of wheat, applesauce, cream cheese, etc.). The judges should be blindfolded or red cellophane installed over the light source in the testing booth so that the color would not influence judgment. Rinse mouth between tasting. In Table 20.3, identify and record each food item in the order that they are presented, and the reason for your selection.

Table 20.3 IDENTIFICATION OF FOOD BY TEXTURE	
Food Item	Reason for Selection
1.	
2.	
3.	
4.	
5.	
6.	
7.	

QUESTIONS

1. What influence does texture have on acceptability and identification of food?

2. What role does sensory evaluation play in rating quality of food?

Glossary of Common Terms Used in Food Preparation

Glossary of Common Terms Used in Food Preparation

Acesulfame-K:	An artificial sweetener that is 200 times sweeter than sugar and, unlike Aspartame, retains sweetness when heated, making it suitable for cooking and baking.
Acid:	(in terms of cooking) Vinegar, lemon juice or cream of tartar.
Active Dry Yeast:	This is the form of tiny dehydrated granules. The yeast cells are alive but dormant because of the lack of moisture. When mixed with warm liquid (105–115°F or 40–46°C) the cells become active. Active dry yeast is available in two forms, *regular* and *quick-rising*, of which the latter takes about half as long to leaven the bread.
Aerate:	A term used in cookery as a synonym for sift.
A La King:	A dish of diced food (usually chicken or turkey) in a rich cream sauce containing mushrooms, pimientos, green peppers, and sometimes sherry.
Al Dente:	A term used to describe pasta or other food that is cooked only until it offers a slight resistance when bitten into, but which is not soft or overdone.
Alkali:	Alkalis counterbalance and neutralize acids. In cooking the most common alkali is bicarbonate of soda, commonly known as baking soda.
Amylopectin:	Highly branched chain fraction of starch involved in the gelatinization process.
Amylose:	Straight chain fraction of starch involved in the gelation process.
Angel Food Cake:	A light, airy sponge-type cake made with stiffly beaten egg whites but no yolks or other fats. It is traditionally baked in a tube pan and is sometimes referred to simply as angel cake.
Antioxidant:	Substance that retards the oxidative process in food, such as fat rancidity and browning of fresh fruits and vegetables, by stopping or interfering with the chain reaction.
Aquaculture:	The cultivation of fish, shellfish, or aquatic plants (such as seaweed) in natural or controlled marine or fresh water environments.
Arborio Rice:	The high starch kernels of this Italian grown grain are shorter and fatter than any other short grain rice.

Artificial Sweeteners:	This category of nonnutritive, high-intensity sugar substitutes include Aspartame, Acesulfame K, Saccharin, and Sucralose.
Arugula:	Also called *rocket*, *roquette*, *rugula*, and *rucola*, arugula is a bitterish, aromatic salad green with a peppery mustard flavor.
Ascorbic Acid (Vitamin C):	Available in powder and tablet form and in mixtures; may be used to prevent darkening of cut or peeled fruits, such as apples, bananas, and peaches.
Aspartame:	An artificial sweetener that is 180–200 times sweeter than sugar.
Bagel:	A doughnut-shaped yeast roll with a dense, chewy texture, and shiny crust. Bagels are boiled in water before they are baked. The water bath reduces starch and creates a chewy crust.
Bake:	To cook in an oven-type appliance. Covered or uncovered containers may be used. When applied to meats in uncovered containers, method is generally called roasting.
Bake Blind:	An English term for baking a pastry shell before it is filled. The shell is usually pricked all over with a fork to prevent it from blistering and rising. Sometimes it is lined with foil or parchment paper, then filled with dried beans or rice, or metal or ceramic pie weights.
Baking Powder:	A leavener containing a combination of baking soda, an acid (such as cream of tartar), and a moisture absorber (such as cornstarch). When mixed with liquid, baking powder releases carbon dioxide gas bubbles that cause a cake or bread to rise. The most common type of baking powder is double-acting which releases some gas when it becomes wet and the rest when exposed to oven heat.
Baking Soda:	Also known as bicarbonate of soda, baking soda is used as a leavener in baked goods. When combined with an acid ingredient such as buttermilk, yogurt, or molasses, baking soda produces carbon dioxide bubbles, thereby causing a dough or batter to rise.
Barbecue:	To roast slowly on a gridiron or spit, over coals, or under free flame, or oven electric unit, usually basting with a highly seasoned sauce. Popularly applied to food cooked in or served with barbecue sauce.
Basmati Rice:	Basmati rice is a fragrant long-grained rice with a fine texture. Its perfumy, nutlike flavor, and aroma can be attributed to the fact that the grain is aged to decrease its moisture content.
Baste:	To moisten meat or other foods while cooking to add flavor, and to prevent drying of the surface. The liquid is usually melted fat, meat drippings, fruit juice, sauce, or water.
Batter:	A mixture of flour and liquid, usually combined with other ingredients as in baked products. The mixture is of such consistency that it may be stirred with a spoon and is thin enough to pour or drop from a spoon.
Bavarian Cream:	A cold dessert composed of a rich custard, whipped cream, various flavorings (fruit puree, chocolate, liqueurs, etc.), and gelatin.
Beat:	To make a mixture smooth by introducing air with a brisk, regular motion that lifts the mixture over and over, or with a rotary motion as with an eggbeater or electric mixer.
Béchamel Sauce:	This French-based white sauce is made by stirring milk into a butter–flour roux (**see Roux**). The thickness of the sauce depends on the proportions of flour and butter to milk.
Biscuit:	In America, biscuits refer to small quick breads, which often use leaveners like baking powder or baking soda. Biscuits are generally savory (but can be sweet) and the texture should be tender and light. In the British Isles, the term "biscuit" usually refers to a flat, thin cookie or cracker.
Blanch:	To precook in boiling water or steam.
Bland:	Mild flavored, not stimulating the taste; smooth; soft-textured.

Blend:	To mix thoroughly two or more ingredients.
Boil:	To cook water or liquid consisting mostly of water in which bubbles rise continually and break on the surface. The boiling temperature of water at sea level is 212°F or 100°C.
Bosc Pear:	A large winter pear with a slender neck and a rusted yellow skin. The Bosc pear holds its shape well when baked or poached.
Braise:	To cook meat or poultry slowly in a covered utensil in a small amount of liquid or in steam.
Bread:	To coat with crumbs of bread or other food; or to coat with crumbs, then with diluted slightly beaten egg or evaporated milk, then again with crumbs.
Brine:	A strong salt solution used in pickling, fermentation, and curing to inhibit growth of certain bacteria and provide flavor.
Broil:	To cook by direct (radiant) heat.
Browning Reactions:	Darkening of some fruits when pared or cut due to oxidation of enzymes. Also, specific protein and sugar reactions.
Bulghur Wheat (Bulgur):	A nutritious staple in the Middle East, bulghur wheat consists of wheat kernels that have been steamed, dried, and crushed. It makes an excellent wheat pilaf.
Butyric Acid:	Found chiefly in butter, this natural acid not only produces butter's distinctive flavor but also causes the rancid smell in spoiled butter.
Caffeine:	An organic compound found in foods such as chocolate, coffee, cola nuts, and tea.
Cake Flour:	A fine-textured, soft-wheat flour with a high starch content.
Candied:	(1) Fruit, fruit peel, or ginger that is cooked in heavy syrup until plump and translucent, then drained and dried. The product is also known as crystallized fruit, fruit peel, or ginger. (2) Sweet potatoes or carrots cooked in sugar or syrup. To candy a food is to cook as described in the preceding.
Canner (Water Bath):	A large, covered cooking utensil with side handles and jar holder. Capacity is designated by the volume of water that the canner will hold. The water capacity must assure a 2- to 4-inch coverage above the tops of the jar.
Caramelize:	To heat sugar or foods containing sugar until a brown color and characteristic flavor develop.
Carbohydrate:	Organic compounds containing carbon, hydrogen, and oxygen; simple sugar (monosaccharides); disaccharides; and polymers of simple sugars (polysaccharides).
Carotenoids:	A variety of yellow to red pigments found in fruits and vegetables that are relatively stable to cooking methods.
Carrageen:	Also called *Irish moss*, carrageen is derived from seaweed. When dried, carrageen is greatly valued as a thickening agent for foods such as puddings, ice cream, and soups.
Casein:	The principle protein in milk, which coagulates with the addition of rennin and its foundation for cheese.
Casserole:	A covered utensil in which food may be baked and served. It may have one or two handles. Size is stated in liquid measurements.
Cellulose:	A polysaccharide found in cell walls of plants, fruits, and vegetables that provide structural rigidity; can be softened by cooking but is not digested in the human alimentary tract.
Chalazae:	The cordlike strands of egg white that are attached to two sides of the yolk, thereby anchoring it in the center of the egg. The more prominent the chalazae, the fresher the egg.
Charlotte:	The classic molded dessert begins with a mold lined with sponge cake, ladyfingers, or buttered bread. The lined mold is then filled with layers (or a mixture) of fruit and custard or whipped cream that has been fortified with gelatin.

Chlorophyll:	The green pigment found in vegetables that becomes olive-green when exposed to an acid-cooking medium.
Chopped:	Cut into pieces with a knife or other sharp tool.
Choux Pastry:	This is also known as cream puff pastry. The dough is created by combining flour with boiling water and butter, then beating eggs into the mixture.
Coagulation:	The change from a fluid state to a thickened curd or clot due to the denaturation of protein.
Cocoa:	The tropical evergreen cacao tree is cultivated for its seeds (also called beans) from which cocoa butter, chocolate, and cocoa powder are produced.
Coddle:	A cooking method most often used with eggs, though other foods can be coddled as well. Coddling is usually done by placing the food in an individual-size container that is covered, set in a large pan of simmering water and placed either on stovetop or in the oven at very low heat.
Colloidal Dispersion:	Combination of small particles and liquid in which the particles are too large to form a true solution and too small to form a coarse suspension; an example of a colloidal dispersion is gelatin and hot water.
Converted Rice:	This rice is also known as parboiled rice. For this rice the unhulled grain has been soaked, pressure steamed, and dried before milling.
Cornstarch:	A dense, powdery "flour" obtained from the endosperm portion of the corn kernel.
Corn Syrup:	A thick, sweet syrup created by processing cornstarch with acids or enzymes.
Cos Lettuce:	This lettuce is also known as Romaine lettuce.
Cream:	To soften a fat, such as shortening or butter, with a fork or other utensil either before or while mixing with another food, usually sugar.
Cream of Tartar:	Cream of tartar is added to candy and frosting mixtures for a creamier consistency, and to egg whites before beating to improve stability and volume.
Creamed:	A term applied to foods that are either cooked or served with a white sauce.
Crepe:	The French word for "pancake," which is exactly what these light, paper-thin creations are.
Crystallization:	Process of forming crystals that result from chemical elements solidifying with an orderly internal structure.
Curdle:	To effect change from a smooth liquid to one in which clots float in a watery medium due to the precipitation of protein either from heat or from acid. Curdling may be observed in milk, cream soups, sauces, custards, and cheese dishes.
Custard:	A pudding-like dessert (made with a sweetened mixture of milk and eggs) that can be baked or stirred on stovetop. Custards require slow cooking and gentle heat in order to prevent separation (curdling).
Cut:	To divide food materials with a knife or scissors.
Cut in:	To distribute solid fat in dry ingredients, such as flour, by chopping with knives or pastry blender until finely divided.
Dash:	Less than 1/8 teaspoon of an ingredient, usually a spice.
Denaturation:	Changing of protein molecule, usually by the unfolding of chains, to a less stable state.
Deep-Fry:	To cook food in hot fat deep enough to completely cover the item being fried.
Devein:	To remove the gray-black vein from the back of the shrimp.

Dextrinization:	Breakdown of starch molecules to dextrins (fragments) by dry heat.
Dextrins:	Segments (fragments) of polysaccharides, such as starch, resulting from the partial hydrolysis caused by either dry heat or acid ingredient.
Dice:	To cut into small cubes.
Disperse:	To distribute or spread throughout some other substance.
Divinity:	A fluffy yet creamy candy made with granulated sugar, corn syrup, and stiffly beaten egg whites.
Dough:	Mixture of flour and liquid, usually with other ingredients added. A dough is thick enough to knead or roll, as in making yeast bread and rolls, but is too stiff to stir or pour.
Dredge:	To cover or coat with flour or other fine substances, such as flour, bread crumbs, or cornmeal.
Dry Measure:	Measuring tool with capacity of 1 cup, 1/2 cup, 1/3 cup, or 1/4 cup. Capacity is based on the relation that 1 cup equals 16 level tablespoons. As the name suggests, the measuring cups are used for measuring dry ingredients, such as flour, sugar, nuts, and cocoa.
Dry Milk:	Milk from which almost all the moisture has been removed. It comes in three basic forms: whole milk, nonfat, and buttermilk.
Durum Wheat:	Wheat that is high in gluten. It is ground into the flour called semolina which is used in the making of pasta.
Emulsification:	A process of breaking up large particles of liquids into smaller ones, which remain suspended in another liquid. Emulsification may be accomplished mechanically, as in the homogenization of ice mixtures; chemically with the use of acid and lecithin (from egg yolk) as in emulsification of oil for mayonnaise; or naturally, in body processes, as when bile salts emulsifying fats during digestion.
Emulsify:	To make into an emulsion. When small drops of one liquid are finely dispersed (distributed) in another liquid, an emulsion is formed. The drops are held in suspension by an emulsifying agent, which surrounds each drop to form a coating.
Enzymatic Browning:	Discoloration found in fruit due to the reaction of enzymes on exposure to oxygen.
Enzyme:	Protein substances that serve as organic catalysts in food affecting changes in color, texture, and flavor; inactivated by exposure to heat; activated slowly by refrigeration or freezing.
Evaporated Milk:	This canned, unsweetened milk is fresh, homogenized milk from which 60% of the water has been removed.
Fat Substitute:	Synthesized substances created to replace fat in a variety of foods.
Fatty Acids:	Organic acids made up of chains of carbon atoms with a carbonyl group on one end; three fatty acids combine with glycerol to make a triglyceride.
Fermentation:	Chemical changes affected by yeast accompanied by the production of alcohol and carbon dioxide.
Fill Weight:	Weight of fruit and vegetables in can as opposed to total weight including liquid.
Finnan Haddie:	Finnan haddie is partially boned, lightly salted, and smoked haddock.
Flaky:	A term describing a food, such as a pie crust, with a dry texture that easily breaks off into flat, flake-like pieces.
Flaxseed:	Though the most universal function of the flaxseed is to produce linseed oil (commonly used in paints, varnishes, linoleums, and inks), this tiny seed contains several essential nutrients including calcium, iron, niacin, phosphorous, and vitamin E. It is also a rich source of Omega-3 fatty acids.

Foam:	A type of colloidal dispersion in which bubbles of gas are surrounded by liquid; specific stability varies.
Fold:	To combine by using two motions; one which cuts vertically through the mixture; the other motion turns it over by sliding the implement across the bottom of the mixing bowl.
Food Additives:	Substances added to a food during preparation. Sometimes a substance is added to increase the concentration of a substance that may be naturally present in the food, such as vitamins. Substances are added to protect the food against spoilage, enhance its flavor, improve its nutritive value, or give it some new property. Additives include chemical preservatives, buffers, and neutralizers, nutrients, non-nutritive sweeteners, coloring agents, stabilizers, emulsifiers, and sequestrants. Some are generally recognized as safe (GRAS); others are allowed for certain foods under certain conditions and in specified amounts. The Food and Drug Administration issues lists of permissible food additives.
Freezer Burn:	Frozen food that has been either improperly wrapped or frozen can suffer from freezer burn—a loss of moisture that affects both texture and flavor. The food will have a dry surface that will look gray or white.
Fricassee:	To cook by braising. Usually applied to fowl, rabbit, or veal cut into pieces.
Fry:	To cook in fat. Applied especially to (1) cooking in a small amount of fat, also called sauté or pan-fry; and (2) cooking in deep layer of fat, also called deep-fat frying.
Fudge:	A creamy semisoft candy most often made with sugar, butter or cream, corn syrup, and various flavors. Fudge is classified as a crystalline candy that undergoes heating, cooling, and beating during its preparation.
Gel:	A liquid in a solid colloidal system that lacks the ability to flow; can be formed by gelatin, pectin, starch, soured milk, and egg.
Gelatin:	An odorless, tasteless, and colorless thickening agent which when dissolved in hot water and then cooled forms a jelly. Gelatin is pure protein derived from beef and pork bones, cartilage, tendons, and other tissue.
Gelatinization:	The absorption of liquid by starch granules accompanied by swelling of the granules and thickening proportional to starch/liquid ratio. Process can proceed while cold, but heat is required to complete the physical change.
Germ:	In the food world, the word "germ" refers to a grain kernel's nucleus or embryo.
Glucose:	A monosaccharide that is often referred to as dextrose. It has about half the sweetening power of regular granulated sugar.
Gluten:	It is the gluten (proteins) in the flour that, when a dough is kneaded, helps hold in the gas bubbles formed by the leavening agent.
Graham Flour:	Whole wheat flour that is slightly coarser than regular grind.
Gravy:	A sauce made from meat juices, usually combined with a liquid such as beef or chicken broth, wine or milk and thickened with flour, cornstarch, or some other thickening agent.
Green Tea:	The tea is produced from leaves that are steamed and dried but not fermented.
Grill:	To cook by direct heat. Also, a utensil or appliance used for such cooking.
Grind:	To reduce to particles by cutting or crushing.
Hard Wheat:	Wheat that is high in protein (10–14%) and yields a flour rich in gluten, making it particularly suitable for yeast bread.
Haricot Vert:	The French term for "green string bean," *haricot* meaning "bean" and *vert* translating as "green."

Hominy:	Hominy is dried white or yellow corn kernels from which the hull and germ have been removed.
Homogenize:	To break up into small particles of the same size. Homogenized milk has been passed through an apparatus to break the fat into small globules that it will not rise to the top as cream. In homogenized shortening, air has been distributed evenly through the fat particles.
Hydration:	Process of absorbing water.
Hydrogenation:	A process in which hydrogen is combined chemically with an unsaturated compound, such as oil, to form solid or semisolid fat.
Hydrolysis:	The process of splitting molecules into simpler components affected by acid, heat, or enzymes.
Immiscible:	Not capable of being mixed.
Infusion:	An infusion is the flavor that is extracted from an ingredient such as tea leaves, herbs, or fruit by steeping them in a liquid (usually hot), such as water, for tea.
Ingredient Labeling:	System that requires food processors to list, on the label, ingredients (in descending order of weight) included in all manufactured items not covered by standards of identity.
International Unit (IU):	Measure of vitamin content, particularly in milk.
Inversion:	Chemical change that sucrose undergoes either when heated with acid or combined with the enzyme invertase in which the molecule is split into its components glucose and fructose.
Irradiation:	A process in which food is exposed to radiation (**see Radiation**).
Jam:	A sweet preserve in which fruit and sugar are cooked together until a thick paste is formed and fruit becomes a homogeneous mass.
Jelly:	A gelled clear product that may be either sweet when made with fruit juice, or savory when made with meat stock.
Julienne:	Meats, fruits, or vegetables cut into slivers resembling matchsticks. Also, a soup with thin strips of vegetables.
Kettle:	A covered or uncovered cooking utensil with a bail handle. Capacity is stated in liquid measurements.
Knead:	To manipulate with a pressing motion accompanied by folding and stretching.
Lactic Acid Bacteria:	Group of microorganisms which converts lactose to lactic acid; responsible for the souring of milk.
Lactose:	This sugar occurs naturally in milk and is also called *milk sugar*. It is the least sweet of all the natural sugars.
Lard:	Rendered and clarified pork fat, the quality of which depends on the area the fat came from and the method of rendering. The very best is *leaf lard* which comes from the fat around the animal's kidneys.
Leavening Agent:	Air, steam, or a microbiological or chemical agent capable of producing carbon dioxide when activated.
Legumes:	Seeds formed in pods, such as peas and beans; the plant has the ability to fix nitrogen in the soil; good source of protein.
Level Off:	To move the level edge of the knife or spatula across the top edge of a container, scraping away the excess material.

Liquid Measure:	Measuring tool with the capacity of 1 quart or less and equipped with a pouring lip for liquids. Capacities and subdivisions include quarts, pints, fluid ounces, or cups. Subdivisions are based on the relation that 1/2 pint equals 1 cup, 236.6 milliliters, or 8 fluid ounces.
Lukewarm:	Approximately 95°F or 35°C; tepid. Lukewarm liquids or foods sprinkled on the wrist will not feel warm.
Maillard Reaction:	Browning reaction involving combination of amino acid from a protein and an aldehyde group from a sugar that leads to the formation of many complex products.
Maltose:	Also called malt sugar, a disaccharide made up of two glucose molecules. Maltose occurs when enzymes react with starches (such as wheat flour) to produce carbon dioxide (which is what makes most bread doughs rise).
Marbling:	Flecks or thin streaks of fat that run throughout a piece of meat, enhancing its flavor, tenderness, and juiciness.
Margarine:	Margarine is made from vegetable oils that are partially hydrogenated. Margarine must contain 80% fat. The other 20% consists of liquid, color, flavor, and other additives.
Marinate:	To let food stand in a marinade which is a liquid, usually an oil–acid mixture.
Mask:	To cover completely. Usually applied to the use of mayonnaise or other thick sauce, but may also be applied to a flavor used as a mask or camouflage flavor.
Melt:	To liquefy by use of heat.
Mince:	To cut or chop into very small pieces.
Mirepoix; Mirepois:	A mixture of diced carrots, onions, celery, and herbs sautéed in butter. Mirepoix is used to season sauces, soups, and stews, as well as for a bed on which to braise foods, usually meats or fish.
Mix:	To combine ingredients in a way that effects a distribution.
Monosodium Glutamate:	A chemical which is added to food to enhance flavor. Its effect on flavor depends on the kinds and amounts of other flavor factors in the food.
Mornay Sauce:	A béchamel sauce to which cheese, usually Parmesan or Swiss, has been added.
Muffin:	A small, cake-like bread that is prepared by adding the liquid ingredients (milk, oil, and egg) to the dry ingredients (flour, sugar, salt, and leavening agent).
Neufchâtel Cheese:	A reduced fat cream cheese.
Nutrition Labeling:	Extensive nutritional information provided on labels of products making a nutritional claim or specifying nutrients added to the product.
Omelet:	A mixture of eggs, seasonings and sometimes water or milk, cooked in butter until firm and filled or topped with various fillings such as cheese, ham, mushrooms, onions, peppers, sausage, and herbs.
Oolong Tea:	Tea that is produced from leaves that are partially fermented, a process that creates teas with a flavor, color, and aroma that fall between black tea and green tea.
Organic Foods:	Foods claimed to be grown without chemical fertilizers or pesticides. As there is no legal definition of this term, it is impossible to verify where used.
Osmotic Pressure:	Force that operates when fruit is simmered, directing the passage of water in or out of the cell, depending upon the surrounding liquid. Also occurs when fruits are sugared and allowed to stand or when a dressing has been held on a green salad for many minutes.
Oven Spring:	The rapid increase in volume of yeast bread during the first few minutes of baking.

Pan-Broil:	To cook, uncovered, on a hot surface, usually in the frypan. Fat is poured off as it accumulates.
Pan-Fry:	To cook in a small amount (thin layer) of fat (**see Fry** and **Sauté**).
Panning:	Method of cooking vegetables in their own juices in a tightly covered pan. A small amount of fat is used to moisten the pan before juices escape.
Panzanella:	An Italian bread salad made with onions, tomatoes, basil, olive oil, vinegar and seasonings, and chunks of bread. Some versions also include cucumbers, anchovies, and/or peppers.
Papain:	An enzyme extracted from papaya and employed as a meat tenderizer.
Parboil:	To boil until partially cooked. Usually cooking is completed by another method.
Pare:	To cut off the outside covering.
Pasteurize:	To preserve food by heating and holding at a specific temperature for a specific length of time sufficient to destroy certain microorganisms and arrest fermentation. Applied to liquids, such as milk and fruit juices. Temperatures used vary with foods but commonly range from 149 to 180°F (60–83°C).
Peel:	To strip off the outside covering.
Pickle:	Method of preserving food employing salt, brine, or vinegar.
Pilaf:	The rice- or bulghur-based dish that begins by first browning the rice in butter or oil before cooking it in stock.
Pinch:	A measuring term referring to the amount of dry ingredient (such as salt or pepper) that can be held between the tips of thumb and the forefinger. It is equivalent to approximately 1/16 teaspoon.
Plasticity:	Ability to be molded or shaped.
Poach:	To cook in a hot liquid using precautions to retain shape. The temperature used varies with the food.
Polyphenols:	Organic compounds with an unsaturated ring and more than one –OH group; implicated in enzymatic browning in foods.
Polyunsaturated Fatty Acid:	Fatty acid that has more than two or more double bonds between carbon atoms.
Pome:	Fruit classification based on central core and presence of seeds; group includes apples, pears, and quinces.
Pot Roast:	A chunky piece of meat cooked by braising (**see Braise**).
Proofing:	The final rising period before baking for yeast doughs that have been molded.
Quiche:	A quiche consists of a pastry shell filled with a savory custard made of eggs, cream, seasonings, and various other ingredients such as onions, mushrooms, ham, shellfish, or herbs.
Quick Bread:	Bread that is quick to make because it does not require kneading or rising time. Examples of quick breads include muffins, biscuits, popovers, and a wide variety of sweet and savory loaf breads.
Radiation:	The combined processes of emission, transmission, and absorption of radiant energy. Radiation is a method of food preservation in which small doses of ionizing radiation are applied to foods in order to prolong the shelf life of perishable foods, such as fresh seafood. Large doses of radiation (2–5 million rads) destroy microbial growth and sterilize the food. For food preservation applications, the radiations available are alpha particles, beta particles, and gamma rays. Quality of the finished product, as well as economic factors, have limited the commercial application of radiation as a means of food preservation.

Rancidity:	State of spoilage unique to fats in which the flavor and odor deteriorate due to either hydrolysis or oxidation.
Rehydration:	To soak, cook, or use other procedures with dehydrated foods to restore water lost during drying.
Render:	To free fat from animal tissues by heating at low temperatures.
Rennet:	Crude extract from calf stomach containing the enzyme rennet.
Retrogradation:	Tendency to gelatinize starch mixtures to form crystalline areas during storage (staling of baked products and starch pastes).
Roast:	To cook uncovered in hot air (dry heat). Meat is oven roasted in an oven or over coals, ceramic briquettes, gas flame, or electric coils. The term is also applied to foods such as corn or potatoes cooked in hot ashes, under coals, or on heated stones or metals.
Roux:	A thickening agent made by heating a blend of flour and fat. It may be white or brown and used in the making of gravies and sauces.
Saturated Fatty Acid:	Fatty acid contains no double bonds between their carbon atoms.
Saturated Solution:	Solution containing all the solute that it can dissolve at that particular temperature.
Sauce:	In the most basic terms, a sauce is a thickened, flavored liquid designed to accompany food in order to enhance and bring out the flavor in food.
Sauté:	To brown or cook in a small amount of fat (**see Fry**).
Scald:	(1) To heat milk to just below the boiling point, when tiny bubbles form at the edge; and (2) to dip certain foods in boiling water (**see Blanch**).
Scallop:	To bake foods (usually cut into pieces) with a sauce or other liquid. The food and sauce may be mixed together or arranged in alternate layers in a baking dish, with or without a topping of crumbs.
Sear:	To brown the surface of meat by a short application of intense heat.
Semolina:	Durum wheat that is more coarsely ground than normal wheat flours, a result that is often obtained by sifting the flour. Most good pasta is made from semolina.
Sherbet:	Sherbet commonly refers to a frozen mixture of sweetened fruit juice and water. It can also contain milk, egg whites, and/or gelatin.
Simmer:	To cook in a liquid just below the boiling point, at temperatures of 185–210°F (85–99°C). Bubbles form slowly and collapse below the surface.
Slurry:	A thin paste of water and flour, which is stirred into hot preparations (such as soups, stews, and sauces) as a thickener. After the slurry is added, the mixture should be stirred and cooked for several minutes in order for the flour mixture to lose its raw taste.
Soft Wheat:	The low protein (6–10%) soft wheat yields flour lower in gluten and therefore better suited for tender baked goods such as biscuits and cakes.
Smoke Point:	Specific point for each fat at which it begins to smoke and emit irritating vapors when heated.
Solution:	A uniform liquid blend containing a solvent (liquid) and a solute (such as sugar) dissolved in the liquid.
Sorbic Acid:	An antimycotic agent used in foods such as cheese to retard mold growth.
Soufflé:	A light, airy mixture that usually begins with a thick egg yolk-based sauce or puree that is lightened by stiffly beaten egg whites.

Sous Vide:	French term for "under vacuum," sous vide is a food-packaging technique pioneered in Europe whereby fresh ingredients are combined into various dishes, vacuum-packed individual pouches, cooked under a vacuum, then chilled.
Steam:	To cook in steam with or without pressure. The steam may be applied directly to the food, as in a steamer or pressure cooker.
Steep:	To allow a substance to stand in liquid below the boiling point for the purpose of extracting flavor, color, or other qualities.
Sterilize:	To destroy microorganisms. Foods are most often sterilized at high temperatures with steam, hot air, or boiling liquid.
Stew:	To simmer food in liquid that covers the food.
Stir:	To mix food materials with a circular motion for the purpose of blending or securing uniform consistency.
Stir-Fry:	Cooking method using tossing motions when cooking over high heat, particularly in Oriental cuisines when a small amount of fat is used.
Sucrose:	A crystalline, water-soluble sugar obtained from sugarcane, sugar beets, and sorghum. It is sweeter than glucose, but not as sweet as fructose.
Supersaturated Solution:	Solution that has dissolved more solute or dispersed substance than it can normally hold at a particular temperature.
Suspension:	Combination of powder and liquid that remains combined only so long as agitation is continued. Solid will settle to bottom when undisturbed (e.g., starch in cold water).
Syneresis:	Drainage of liquid from a gel system when cut or disturbed.
Texture:	Properties of food, including roughness, smoothness, graininess, creaminess, and so on.
Textured Vegetable Protein:	Fabricated meats and meat extenders processed from soybeans providing excellent nutritive qualities and the potential for economy when sufficient volume can be utilized.
Toast:	To brown by means of direct (radiant) heat.
Tofu:	Also known as soybean curd and bean curd, custard-like white tofu is made from curdled soymilk, an iron-rich liquid extracted from ground cooked soybeans.
Trans Fatty Acids:	A type of fat created when oils are hydrogenated, which chemically transform them from their normal liquid state (at room temperature) into solids. Trans fatty acids can be found in a wide array of processed foods, including cookies and margarines.
Translucent:	Shining or glowing through; partly translucent.
Tuber:	Enlarged, fleshy portions of root systems growing underground that can reproduce when the "eyes" are planted (e.g., potatoes).
Variety Meat:	Variety meats are the animal innards and extremities that can be used in cooking. They include brains, feet, heart, kidneys, liver, sweetbreads, tongue, and tripe.
Vegetable Oil:	Any of various edible oils made from a plant source such as vegetables, seeds, or nuts.
Velouté Sauce:	One of the five "mother sauces," velouté is a stock-based white sauce. It can be made from chicken or veal stock or fish fumet (concentrated stock) thickened with a white roux.
Viscosity:	A property of fluids that determines whether they flow readily or resist flow. A pure liquid at a given temperature and pressure has a definite viscosity, which usually increases with a decrease in temperature. Sugar syrups, for example, thicken as their temperatures decrease.
Whey:	Liquid portion of milk remaining after the curd, which is chiefly the protein casein, is precipitated.

White Chocolate:	Not really chocolate at all, white chocolate is typically a mixture of sugar, cocoa butter, milk solids, lecithin, and vanilla.
Whip:	To beat rapidly such mixtures as gelatin dishes, eggs, and cream to incorporate air and increase volume.
Yeast:	Yeast is a living, microscopic, single-cell organism that, as it grows, converts its food (through the process known as **fermentation**) into alcohol and carbon dioxide. This trait is important when making bread.
Yogurt:	A dairy product that is the result of milk that has fermented and coagulated because it is invaded by friendly bacteria. This can be accomplished naturally by keeping the milk at about 110°F (43°C) for several hours.
Zest:	The aromatic outermost skin layer of citrus fruit (usually oranges and lemons), which is removed with the aid of citrus zester, paring knife, or microplane. The aromatic oils in citrus zest are what add flavor to food.

APPENDIX A

Food Safety

APPENDIX A
Food Safety

WHAT IS A FOODBORNE ILLNESS?

1. A foodborne illness is a disease that is transmitted to people through food.
2. Foodborne illnesses are caused by microorganisms, such as include bacteria, viruses, parasites, and fungi.
3. Food that allows microorganisms to grow and requires **time–temperature control** for safety is called potentially hazardous food.

WHAT CONSTITUTES POTENTIALLY HAZARDOUS FOODS?

1. Milk and milk products
2. Sliced melon
3. Untreated garlic-and-oil mixtures
4. Beef, pork, and lamb
5. Poultry
6. Shellfish and crustaceans
7. Fish
8. Raw sprouts and sprout seeds
9. Baked potatoes
10. Eggs (Except those treated to eliminate Salmonella spp.)
11. Tofu or other soy-protein food
12. Cooked rice, beans, and vegetables

HOW CAN FOOD BECOME UNSAFE?

1. Time–temperature abuse: A food can become contaminated when it has been allowed to stay in the **Danger Zone: between 41 and 135°F (5 and 57°C).**
2. Cross-contamination: It occurs when microorganisms are transferred from one food or surface to another. **Prevent Cross-Contamination by** handwashing; proper cleaning and sanitizing; and proper storage of raw food below ready-to-eat food.
3. Poor personal hygiene: People who do not wash their hands properly, or often enough, are the biggest risks to food safety.
4. Improper cleaning and sanitizing: Even though proper preparation is followed, without proper cleaning and sanitizing, people can become ill.

WHAT ARE THE FOUR ACCEPTABLE WAYS TO THAW FOOD?

1. Thaw food in the refrigerator at 41°F (5°C) or lower.
2. Thaw food by submerging it under drinkable, running water, that is 70°F (21°C) or lower.
3. Thaw food in the microwave only if the food will be cooked immediately.
4. Thaw food as part of the cooking process.

WHAT ARE THE MINIMAL INTERNAL COOKING TEMPERATURES FOR SOME COMMON FOODS?

Food Item	Time and Temperature
Poultry (whole or ground)	165°F(74°C) for 15 seconds
Ground meat/ground fish	155°F(68°C) for 15 seconds
Pork and beef	145°F(63°C) for 15 seconds
Fish	145°F(63°C) for 15 seconds

WHAT IS THE DANGER ZONE?

If potentially hazardous food is not held at the proper temperature, microorganisms present in the food can grow and make someone ill. Food must be kept out of the **Danger Zone** when being held for service:

Hold cold food at 41°F (5°C) or lower ←Temperature Danger Zone→ Hold hot food at 135°F (57°C) or higher

APPENDIX B

Sanitation in the Kitchen

Appendix B
Sanitation in the Kitchen

The objective is to remove food particles as well as other soil, and to control bacteria. All equipment and utensils must be expected to contain food spoilage bacteria, as well as pathogens (disease causing bacteria). Proper cleaning removes soil. Sanitizing reduces the bacterial load to a safe level. Proper water temperature and proper amounts of cleaning compound must be used for an effectual job. It is essential to use cleaning compound or detergent along with a brush (mechanical or hand activated) to loosen food remains. A thorough rinsing must follow in order to sanitize the surfaces to achieve the desired bacterial action.

To be effective, cleaning and sanitizing must be a **two-step process:**

1. Surface must be cleaned first.
2. Then the surface is rinsed and sanitized.

Therefore, any surface that comes in contact with food must be **cleaned, rinsed, and sanitized:**

- Each time you use it.
- When you are interrupted during a task.
- When you begin working with another type of food.
- As often as possible, but at least every 4 hours if you are using something constantly.

General rules that must be followed during preparation of food in the kitchen:

1. All raw fruit and vegetables must be washed before being cooked or served.
2. Cold food should be kept cold, and hot food should be kept hot.
3. Get food hot as quickly as possible and keep it hot; above 135°F (57°C).
4. Get food cold as quickly as possible and keep food cold at 41°F (5°C) or below.
5. Do not expose food to the Danger Zone (41°F [5°C] to 135°F [57°C]) for more than 2 hours.
6. Separate surfaces (cutting boards) should be used for the cutting, cubing, and portioning of raw meats and poultry, and for cooked meat items.
7. Each time a person has had contact with uncooked items which may be suspected of carrying pathogens, he or she must wash his or her hands before handling cooked items.
8. Always sanitize your work area before beginning any type of food preparation. Use a solution of bleach (1 teaspoon) and water (1 quart).
9. If in doubt about the safety or quality of the food, throw it out.
10. Keep food covered as much as possible; use clean utensils.

WASHING DISHES

Hand dishwashing should follow the prescribed directions:

1. Fill sink or pan with hot water (120°F [49°C] or above). Add enough detergent to make light suds.
2. Rinse dishes and utensils in clean hot water.
3. Avoid towel drying your dishes, glasses, or utensils.
4. Use any recommended soap or detergent that will be adequate in cleaning as well as sanitizing.

MACHINE DISHWASHING

1. Remove all food particles from dishes, using either a scraper or the rinse water power arm.
2. Pre-rinse dishes at 80°F (27°C). Wash at 140°F (60°C) and rinse at 180°F (82°C). Avoid toweling dishes.
3. Store dishes in a clean, dry, enclosed storage area. Invert glasses and cups.
4. Use any recommended soap or detergent that will be adequate in cleaning as well as sanitizing.

APPENDIX C

Care and Cleaning of Small Applicances

Appendix C
Care and Cleaning of Small Applicances

BLENDERS

1. Unplug; remove blender jar; remove all parts and place in **hot soapy water**. Use a dish cloth to wash parts. Rinse in hot water and drain.
2. Use a damp (not wet) cloth to wipe blender base. Wrap cord around top.
3. When parts are dry **do not assemble**; blades need to air dry longer.
4. Return blender to storage area.

HAND MIXERS

1. Unplug; remove beaters and place in **hot soapy water**. Wash thoroughly. Rinse in hot water and drain.
2. Use a damp (not wet) cloth to wipe the mixer. Wrap cord around.
3. When beaters are dry, replace with mixer. Return to storage area.

STAND MIXERS

1. Unplug; remove beater and place in **hot soapy water**.
2. Wash thoroughly and rinse in hot water; drain.
3. Rinse out mixing bowl with warm water. Then use hot soapy water to clean the bowl thoroughly. Rinse in hot water; drain.
4. Use a damp (not wet) cloth to wipe mixer. Wrap cord around neck.
5. Replace clean bowl onto mixing stand and place the clean beater(s) into the bowl.

SIFTERS

1. **DO NOT WASH!**
2. Tap the sifter gently over the sink to remove any excess flour trapped on the sides.
3. Use a damp (**not wet**) cloth or paper towel to wipe all trace of flour or sometimes cocoa (if being used).
4. Allow the sifter to air dry before storing.

DEEP FAT FRYERS

1. Unplug; leave on counter to allow the fat to cool.
2. Pour fat into a container; cover tightly.
3. Follow manufacturer's directions for cleaning the fryer. Never immerse the main part of the fryer into water. Use hot soapy water to cut the fat and grease.
4. Use a scouring soap pad to clean the fry basket, especially if there is a sticky build-up of fat on the basket.

APPENDIX D

Measuring Equivalents

APPENDIX D
Measuring Equivalents

TABLE OF EQUIVALENTS		
Pinch or dash	=	less than 1/8 teaspoon
3 teaspoons	=	1 tablespoon
2 tablespoons	=	1 fluid ounce
1 jigger	=	1½ fluid ounces
4 tablespoons	=	1/4 cup
5 tablespoons + 1 teaspoon	=	1/3 cup
8 tablespoons	=	1/2 cup
10 tablespoons + 2 teaspoons	=	2/3 cup
12 tablespoons	=	3/4 cup
16 tablespoons	=	1 cup

SOME FRACTIONAL MEASURES		
1/2 of 1/4 cup	=	2 tablespoons
1/2 of 1/3 cup	=	2 tablespoons + 2 teaspoons
1/2 of 1/2 cup	=	1/4 cup
1/2 of 2/3 cup	=	1/3 cup
1/2 of 3/4 cup	=	1/4 cup + 2 tablespoons
1/3 of 1/4 cup	=	1 tablespoon + 1 teaspoon
1/3 of 1/3 cup	=	1 tablespoon + 2⅓ teaspoons
1/3 of 1/2 cup	=	2 tablespoons + 2 teaspoons
1/3 of 2/3 cup	=	3 tablespoons + 1⅔ teaspoons
1/3 of 3/4 cup	=	1/4 cup

LIQUID MEASURES			
Fluid Ounces	U.S.	Imperial	Milliliters
	1 teaspoon	1 teaspoon	5
1/4	2 teaspoons	1 dessert teaspoon	10
1/2	1 tablespoon	1 tablespoon	14
1	2 tablespoons	2 tablespoons	28
2	1/4 cup	4 tablespoons	56
4	1/2 cup		110
5		1/4 pint/1 gill	140
6	3/4 cup		170
8	1 cup		225
9			250, 1/4 liter
10	1¼ cups	1/2 pint	280
15		3/4 pint	420
16	2 cups		450
18	2¼ cups		500, 1/2 liter
20	2½ cups	1 pint	560
24	3 cups		675
25		1¼ pints	700
27	3½ cups		750
30	3¾ cups	1½ pints	840
32	4 cups		900
36	4½ cups		1000, 1 liter
40	5 cups	2 pints	1120

SOLIDS MEASURES			
U.S. and Imperial Measures		Metric Measures	
Ounces	Pounds	Grams	Kilos
1		28	
2		56	
3½		100	
4	1/4	112	
5		140	
6		168	
8	1/2	225	
9		250	1/4
12	3/4	340	
16	1	450	
18		500	1/2
20	1¼	560	
24	1½	675	
27		750	3/4
32	2	900	
36	2¼	1000	1

OVEN TEMPERATURE EQUIVALENTS			
Fahrenheit	Celsius	Gas Mark*	Description
225	107	1/4	Cool
250	121	1/2	
275	135	1	Very slow
300	148	2	
325	163	3	Slow
350	177	4	Moderate
375	191	5	
400	204	6	Moderately hot
425	218	7	Fairly hot
450	232	8	Hot
475	246	9	Very hot
500	260	10	Extremely hot

*Gas Mark: A system and unit of marking temperatures on gas ovens and cookers in the United Kingdom and Commonwealth of Nations countries.

APPENDIX E

Emergency Substitutions

APPENDIX E
Emergency Substitutions

In a pinch, any of the following ingredient substitutions can be used successfully except in temperamental cakes, breads, cookies, or pastries.

LEAVENING

- $1\frac{1}{2}$ teaspoons phosphate or tartrate baking powder = 1 teaspoon double-acting baking powder
- 1/4 teaspoon baking soda + 1/2 teaspoon cream of tartar = 1 teaspoon double-acting baking powder
- 1/4 teaspoon baking soda + 1/2 cup sour milk = 1 teaspoon double-acting baking powder in liquid mixtures; reduce recipe liquid content by 1/2 cup

THICKENING

- 1 tablespoon cornstarch = 2 tablespoons all-purpose flour
- 1 tablespoon potato flour = 2 tablespoons all-purpose flour
- 1 tablespoon arrowroot = $2\frac{1}{2}$ tablespoons all-purpose flour
- 2 teaspoons quick-cooking tapioca = 1 tablespoon all-purpose flour (use in soups only)

SWEETENING, FLAVORING

- 1 cup granulated sugar + 2 tablespoons molasses or dark corn syrup = 1 cup dark brown sugar, packed
- $1\frac{1}{4}$ cups granulated sugar + 1/3 cup liquid = 1 cup light corn syrup or honey
- 3/4 cup light corn syrup + 1/4 cup molasses = 1 cup dark corn syrup
- 3 tablespoons cocoa + 1 tablespoon butter = 1 (1 ounce) square unsweetened chocolate
- 1 tablespoon baking cocoa + 2 teaspoons granulated sugar + 2 teaspoons shortening = 1 ounce of semisweet baking chocolate
- 1/8 teaspoon cayenne pepper = 3 to 4 drops liquid hot red pepper seasoning

FLOUR

- 1 cup all-purpose flour minus 2 tablespoons = 1 cup cake flour
- 1 cup + 2 tablespoons cake flour = 1 cup all-purpose flour
- 1 cup self-rising flour = 1 cup all-purpose flour + $1\frac{1}{4}$ teaspoons baking powder and a pinch of salt

DAIRY

- 1/2 cup evaporated milk + 1/2 cup water = 1 cup whole milk
- 1 cup skim milk + 2 teaspoons melted butter = 1 cup whole milk
- 1 cup whole milk + 1 tablespoon lemon juice or white vinegar = 1 cup buttermilk (or sour milk). Let it stand for 5–10 minutes before using
- 3/4 cup milk + 1/4 cup melted butter = 1 cup light cream

EGGS

- 2 egg yolks = 1 egg (for thickening sauces, custards)
- 2 egg yolks + 1 tablespoon cold water = 1 egg (for baking)
- $1\frac{1}{2}$ tablespoons stirred egg yolks = 1 egg yolk
- 2 tablespoons stirred egg whites = 1 egg white
- 3 tablespoons mixed broken yolks and whites = 1 medium size egg
- 1/4 cup fat-free egg product = 1 egg

221

MISCELLANEOUS

- 1 cup boiling water + 1 bouillon cube or envelope instant broth mix = 1 cup broth
- 1 teaspoon beef extract blended with 1 cup boiling water = 1 cup beef broth
- 1 cup fine bread crumbs = 3/4 cup fine cracker crumbs
- 1/2 cup minced, plumped, pitted prunes or dates = 1/2 cup seedless raisins or dried currants
- 6 tablespoons mayonnaise blended with 2 tablespoons minced pickles or pickle relish = 1/2 cup tartar sauce
- 3/4 to 1 teaspoon dried herbs = 1 tablespoon herbs, chopped fresh
- 1/2 teaspoon ground mustard + 2 teaspoons cider or white vinegar = 1 tablespoon yellow mustard
- 3/4 teaspoon ground sage + 1/4 teaspoon ground thyme = 1 teaspoon poultry seasoning
- 1/2 teaspoon ground cinnamon + 1/4 teaspoon ground ginger + 1/8 teaspoon ground allspice + 1/8 teaspoon ground nutmeg = 1 teaspoon pumpkin or apple pie spice
- 1/2 cup tomato sauce + 1/2 cup water = 1 cup tomato juice
- 1 cup tomato sauce cooked uncovered until reduced to 1/2 cup = 1/2 cup tomato paste

APPENDIX F

Safe Food Storage

APPENDIX F
Safe Food Storage

REFRIGERATOR AND FREEZER FOOD STORAGE		
Foods	34–40°F (Refrigerator)	0°F or below (Freezer)
BAKED PRODUCTS		
Breads—muffins, quick breads, yeast breads	5–7 days	2–3 months
Cakes—unfrosted and frosted	3–5 days	Unfrosted: 3–4 months Frosted: 2–3 months
Cheesecakes—baked	3–5 days	4–5 months
Cookies—baked	If in recipe	Unfrosted: 12 months Frosted: 3 months
Pies—unbaked or baked	3–5 days	Unbaked fruit pie, fruit pies, baked pecan: 2–3 months
VEGETABLES AND FRUITS		
Vegetables—commercially frozen		8 months
Vegetables—home frozen		12 months
Fruits—commercially frozen		12 months
Fruits—home frozen		12 months

SAFE STORAGE IN PANTRY

CANNED GOODS	STORAGE TIME
Fruit	1 year
Vegetables	1 year
Soup	1 year
Meat, fish, poultry	1 year
PACKAGED MIXES	
Cake mix	1 year
Casserole mix	18 months
Frosting mix	8 months
Pancake mix	6 months
STAPLES	
Baking powder and soda	1 year
Breakfast cereal:	
Ready-to-eat	Check package date
Uncooked	1 year
Coffee (opened and refrigerated)	6–8 months
Cornmeal: regular and self-rising	10 months
Dried beans and peas	18 months
Flour:	
All-purpose	10–15 months
Whole wheat (refrigerated)	3 months
Grits:	
Regular	10 months
Instant	9 months

(continued)

CANNED GOODS	STORAGE TIME
Milk: evaporated and condensed	1 year
Pasta	10–15 months
Peanut butter	6 months
Salt, pepper, sugar	18 months
Shortening	8 months
Spices:	
Ground	6 months
Whole (discard if aroma fades)	1 year
Tea bags	1 year
Vegetable oil	3 months
Worchestershire sauce	2 years

APPENDIX G

Retail Meat Cuts and Recommended Cooking Methods

Beef Made Easy®
Retail Beef Cuts and Recommended Cooking Methods

Rib — **Loin**
Chuck — **Sirloin**
Shank — **Round**
Brisket   Plate   Flank

BEEF
IT'S WHAT'S FOR DINNER.

BEEF
FUNDED BY
THE BEEF CHECKOFF

Chuck

| CHUCK 7-BONE POT ROAST | CHUCK POT ROAST Boneless | CHUCK STEAK Boneless | CHUCK EYE STEAK Boneless | SHOULDER TOP BLADE STEAK | SHOULDER TOP BLADE STEAK Flat Iron |

| SHOULDER POT ROAST * Boneless | SHOULDER STEAK * Boneless | SHOULDER CENTER * Ranch Steak | SHOULDER PETITE TENDER * | SHOULDER PETITE TENDER MEDALLIONS * | BONELESS SHORT RIBS |

Rib

| RIB ROAST | RIB STEAK | RIBEYE ROAST Boneless | RIBEYE STEAK Boneless | BACK RIBS |

Loin

| PORTERHOUSE STEAK | T-BONE STEAK * | TOP LOIN STEAK * Bone-in | TOP LOIN STEAK * Boneless | TENDERLOIN ROAST * | TENDERLOIN STEAK * |

Sirloin

| TRI-TIP ROAST * | TRI-TIP STEAK * | TOP SIRLOIN STEAK * Boneless |

Key to Recommended Cooking Methods

- Skillet
- Grill or Broil
- Marinate & Grill or Broil
- Stir-Fry
- Roast
- Stew
- Braise
- Pot Roast

Round

| TOP ROUND STEAK * | BOTTOM ROUND ROAST * | BOTTOM ROUND STEAK * Western Griller | EYE ROUND ROAST * | EYE ROUND STEAK * |

| ROUND TIP ROAST * | ROUND TIP STEAK * | SIRLOIN TIP CENTER ROAST * | SIRLOIN TIP CENTER STEAK * | SIRLOIN TIP SIDE STEAK * |

Shank and Brisket

| SHANK CROSS CUT * | BRISKET FLAT CUT * |

Plate and Flank

| SKIRT STEAK | FLANK STEAK * |

* These cuts meet government guidelines for "lean" and are based on cooked servings with visible fat trimmed.

Lean is defined as less than 10 grams of total fat, 4.5 grams of saturated fat, and less than 95 milligrams of cholesterol per serving and per 100 grams (3.5 oz).

Other

| GROUND BEEF | CUBED STEAK | BEEF FOR STEW | BEEF FOR KABOBS | BEEF FOR STIR-FRY OR FAJITAS |

©2006 CATTLEMEN'S BEEF BOARD AND NATIONAL CATTLEMEN'S BEEF ASSOCIATION   10503   1006

Reprinted through the permission of The Cattleman's Beef Board and The National Cattleman's Beef Association.

228

CHICKEN
Foodservice Cuts

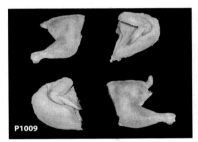

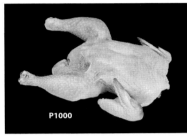

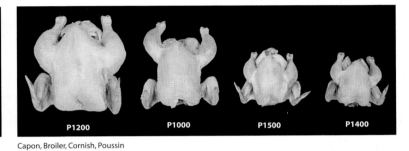

P1000

Broiler

P1200   P1000   P1500   P1400

Capon, Broiler, Cornish, Poussin

P1009

Broiler, Quartered

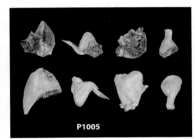

P1005

Eight-Piece Broiler

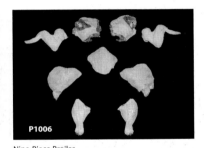

P1006

Nine-Piece Broiler

P1036
P1037
P1038

Wings

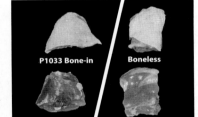

P1033 Bone-in   Boneless

Thighs

Boneless Fillet   Butterflied

Skin-on Butterflied

Boneless Breasts

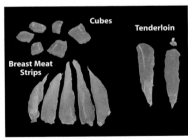

Cubes   Tenderloin

Breast Meat Strips

Breast Meat

Portion-Controlled Breast Meat

Pulled Meat

Breaded Tenderloin

Breaded Nuggets   Diced Chicken

Value-Added Chicken

NAMP The Meat Buyer's Guide

PORK

Foodservice Cuts

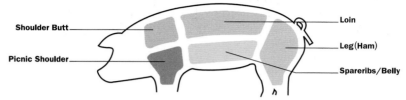

Shoulder Butt — Loin

Picnic Shoulder — Leg (Ham)

Spareribs/Belly

NAMPS/IMPS Number (North American Meat Processors Association/Institutional Meat Purchase Specifications)
©1997 North American Meat Processors Association

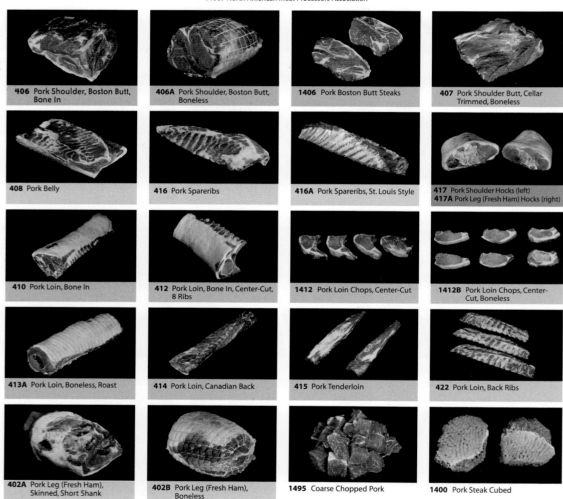

406 Pork Shoulder, Boston Butt, Bone In

406A Pork Shoulder, Boston Butt, Boneless

1406 Pork Boston Butt Steaks

407 Pork Shoulder Butt, Cellar Trimmed, Boneless

408 Pork Belly

416 Pork Spareribs

416A Pork Spareribs, St. Louis Style

417 Pork Shoulder Hocks (left)
417A Pork Leg (Fresh Ham) Hocks (right)

410 Pork Loin, Bone In

412 Pork Loin, Bone In, Center-Cut, 8 Ribs

1412 Pork Loin Chops, Center-Cut

1412B Pork Loin Chops, Center-Cut, Boneless

413A Pork Loin, Boneless, Roast

414 Pork Loin, Canadian Back

415 Pork Tenderloin

422 Pork Loin, Back Ribs

402A Pork Leg (Fresh Ham), Skinned, Short Shank

402B Pork Leg (Fresh Ham), Boneless

1495 Coarse Chopped Pork

1400 Pork Steak Cubed

The above cuts are a partial representation of NAMPS/IMPS items.

NAMP The Meat Buyer's Guide

APPENDIX H

How to Identify Cuts of Meat by Bones

TYPE OF BONE	COMMON NAME	GENERAL COOKING METHOD
ARM	beef chuck and steak	braise, cook in liquid
	beef chuck arm pot roast	braise, cook in liquid
	lamb shoulder arm chop	braise, broil, panfry
	lamb shoulder arm roast	roast
	pork shoulder arm steak	braise, panfry
	pork shoulder arm roast	roast
	veal shoulder arm steak	braise, panfry
	veal shoulder arm roast	braise, roast
BLADE (center cuts)	beef chuck blade steak	braise, cook in liquid
	beef chuck blade pot roast	braise, cook in liquid
	lamb shoulder blade chop	braise, broil, panfry
	lamb shoulder blade roast	roast
	pork shoulder blade steak	braise, broil, panfry
	pork shoulder blade Boston roast	braise, roast
	veal shoulder blade steak	braise, panfry
	veal shoulder blade roast	braise, roast
RIB (backbone and rib bone)	beef rib steak	broil, panfry
	beef rib roast	roast
	lamb rib chop	broil, panfry
	lamb rib roast	roast
	pork rib chop	braise, broil, panfry
	pork rib roast	roast
	veal rib chop	braise, panfry
	veal rib roast	roast

TYPE OF BONE	COMMON NAME	GENERAL COOKING METHOD
LOIN (backbone; T-shape)	beef loin steak (T-bone, porterhouse) beef loin tenderloin roast or steak lamb loin chop lamb loin roast pork loin chop pork loin roast veal loin chop veal loin roast	broil, panfry roast, broil broil, panfry roast braise, broil, panfry roast braise, panfry roast, braise
HIP - Pin bone (near short loin)	beef sirloin steak beef loin, tenderloin, roast, or steak	broil, panfry roast, broil
Flat bone (center cut)	lamb sirloin chop lamb leg roast pork sirloin chop pork sirloin roast	broil, panfry roast braise, broil, panfry roast
Wedge bone (near round)	veal sirloin steak veal leg sirloin roast	braise, panfry roast

TYPE OF BONE	COMMON NAME	GENERAL COOKING METHOD
LEG (leg or round bone)	beef round steak beef rump roast	braise, panfry braise, roast
	lamb leg steak lamb leg roast	broil, panfry roast
	pork leg (ham) steak pork leg roast (fresh or smoked)	braise, broil, panfry roast
	veal leg round steak veal leg roast	braise, panfry braise, roast
BREAST (breast and rib)	beef brisket (fresh or corned) beef plate short rib	braise, cook in liquid braise, cook in liquid
	lamb breast lamb breast riblet	roast, braise braise, cook in liquid
	pork bacon (side pork) pork sparerib	broil, panfry, bake roast, braise, cook in liquid
	veal breast veal breast riblet	roast, braise braise, cook in liquid